PARTICIPATORY IRRIGATION MANAGEMENT

PARTICIPATORY IRRIGATION MANAGEMENT

Evolution, Perception and Impact

A. Narayanamoorthy
S.S. Kalamkar

GYAN

Participatory Irrigation Management: Evolution, Perception and Impact

ISBN: 978-81-212-1094-2

Published in 2011 in India by
Gyan Publishing House
23, Main Ansari Road, Darya Ganj,
New Delhi-110002
Phones : 23282060, 23261060
Fax : 011-23285914
E-mail: books@gyanbooks.com
Website: www.gyanbooks.com

Leser Type Setting by : Square Vision, Delhi

Printed at : Chawla Offset Press, Delhi

CONTENTS

CONTENTS

LIST OF TABLES

FOREWORD

The irrigation sector of India is one of the largest in the world, but the overall performance of irrigation sector has been deteriorating over the years. While the water use efficiency is very low under surface method of irrigation, the revenue generated from the publicly funded irrigation sector is not even enough to manage the operation and maintenance expenditures of the sector. Poor participation of farmers in irrigation management related activities is cited as one of the major reasons for the deteriorating performance of the irrigation sector. Therefore, efforts have been made to transfer the system management from the control of State to Water Users' Associations (WUAs) in recent years. In Maharashtra State, which has the largest irrigation sector in India, a significant policy decision has been made to handover the irrigation management to Water Users Association during 2001 and an Act namely Maharashtra Management of Irrigation System by Farmers Act: 2005 has also been enacted to support the same. Though studies on the participatory irrigation management are available in India, not many attempts have been made to study the varied impact of WUAs on different parameters that are established in different locations using different sources of water. In this study, using field data collected from 150 sample farmers selected from 10 different WUAs, Dr. A. Narayanamoorthy and Dr. S. S. Kalamkar made an in-depth analysis on the impact of participatory irrigation management on various parameters. It is hoped that the findings and policy suggestions of the study would be useful to the policy makers and researchers concerning participatory irrigation management in our country.

R.S. Deshpande
Director
Institute for Social and Economic Change

January 5, 2011 Bangalore - 560 072

PREFACE

In view of the importance of irrigation in the overall development of agriculture and rural economy, Indian policy makers have been giving paramount thrust to the development of irrigation since independence. This has made a significant impact to the area under irrigation, increasing from about 26 million hectares in 1950-51 to over 84 mha in 2005-07. Today, India has the largest irrigation sector in the world. However, the irrigation sector of India currently has been facing a number of different problems. Apart from poor water use efficiency under surface method of irrigation, the financial recovery of State managed irrigation sector is not even enough to manage the operation and maintenance expenditures of the sector. The experiences of different countries show that the poor performance of the irrigation sector can be improved substantially by transferring the operation and management of the system from State control to users. Thus, increased importance has been given to involve the participation of farmers in irrigation management since the late-eighties. In Maharashtra State, participation of farmers in irrigation management activities has been taken up seriously since the mid-nineties. While considerable progress has been made to handover the irrigation management to Water Users' Associations (WUAs) in the State, not many attempts have been made to study as to how the heterogeneity in the physical characteristics (location and source of water used) of WUAs affect the performance of the same. In this study, using field data collected from 150 sample farmers selected from 10 different WUAs in Maharashtra State, an attempt has been made to evaluate the overall performance of the participatory irrigation management. This book is extensively modified version of the study which was originally initiated at the instance of the Ministry of Agriculture, Government of India, New Delhi and was carried out at the Agro-Economic Research Centre, Gokhale Institute of Politics and Economics (GIPE), Pune, Maharashtra.

This study has benefited from various scholars and officials at different levels and times. First of all, we would like to thank the former Director of GIPE, Prof. V.S. Chitre, with whom we have had initial discussions on the study that helped us to refine the focus of the study. Prof. Vasant P. Gandhi and Dr. N.V. Namboodiri, Centre for Management in Agriculture, Indian Institute of Management, Ahmedabad, not only provided suggestions at different times of the study but also have gone through the draft report of the study and offered comments. Our discussions with Dr. R.K. Patil, Er. S.N. Lele and Mr. K.J. Joy of Society for Promoting Participative Ecosystem Management (SOPPECOM), Pune, enriched our knowledge on the participatory irrigation management in Maharashtra. We are grateful to all of them for their suggestions and support for the study.

A number of government officials associated with Water Resources Department, Government of Maharashtra, helped us in carrying out the study. Mr. V. M. Ranade, former Secretary, Water Resources Department, Government of Maharashtra, not only provided valuable inputs on the participatory irrigation management but also helped us in getting secondary data required for the study. Mr. Saha, Director, Directorate of Irrigation Research and Development (DIRD), Government of Maharashtra, Pune, provided all the required data on the development of participatory irrigation management in the State. While selecting WUAs at different locations for the study, we have benefited from Superintendent Engineer, Command Area Development Agency (CADA), Sinchan Bhavan, Pune, Executive Engineer, Kukadi Irrigation Division and other officials of Sinchan Bhavan, Pune. We would like to record our sincere thanks to all the government officials for their invaluable help.

We have also received support and encouragements from some of our colleagues in the Gokhale Institute of Politics and Economics while carrying out the study. We would specifically thank the Director, Officiating Director, Officer-in-Charge of Agro-Economic Research Centre and Library staff of Gokhale Institute of Politics and Economics for their support. We are also thankful to Mrs. Shubhangi A Khuperkar and Mrs. Manu Jain for processing the data using computer and Mr. R.D. Khodaskar, Mr. S.S. Dete, Mr. V.G. Kasbe and Mr. V.B. Lokare for collecting data from field and other sources.

The study would not have reached to this stage without the active co-operation of the office bearers of the selected WUAs and the sample farmers, who provided all the required data for the study without any hesitation and expectation. We thank each one of them for their invaluable support. Finally, we would like to underline that none of the above mentioned scholars, individuals and institutions are responsible for the errors remaining in the study.

A. Narayanamoorthy

January 5, 2011 **S.S. Kalamkar**

1

INTRODUCTION

Water plays a paramount role in every aspect of human life. In agriculture, irrigation water is always considered to be an engine of agricultural growth. Irrigation development increases the cropping intensity, alters the cropping pattern in favour of high value crops, encourages the adoption of technological inputs (HYV seeds, fertilisers, pesticides, etc.) as well as machineries. All of these help to augment the crop output one way or the other. Besides providing direct benefits to the farming community, Irrigation Development also indirectly benefits the non-farming community substantially. While benefiting the landless agricultural labourers in terms of increased employment opportunities and wage rate, it helps to reduce the rural poverty in a sustained manner (Narayanamoorthy, 2001; Saleth, et al., 2003; Hussain and Hanjra, 2003; Bhattarai and Narayanamoorthy, 2003; Shah and Singh, 2004.) Increased production of foodgrains and other commodities that takes place mainly because of irrigation development also makes dent in the prices of agricultural commodities, which indirectly benefits millions of non-agricultural rural and urban consumers.

Investment in Irrigation

In view of the importance of irrigation development in agriculture and other sectors, the policy makers in India have been giving great priority to the development of irrigation since independence. The investment in irrigation sector has increased manifold over the years, from Rs. 455.07 crore in the first plan (1951-56) to a level of Rs. 63681.55 crore at the end of the ninth plan (1997-2002) (GOI, 2002). Altogether about Rs. 1556.25 billion (in current prices) have been spent exclusively for the development of irrigation by the public sector alone at the end of

2001-02[1]. As a result of massive investment, the area under irrigation has increased from 26.61 million hectares (mha) in 1950-51 to over 80 mha in 2001-02. The coverage of irrigation to cropped area increased from 17.4 per cent in 1950-51 to over 42 per cent in 2003-04. This massive irrigation development has also increased the cropping intensity from about 111 per cent in 1950-51 to over 134 per cent in 2002-03 at the all India level.

Despite significant achievements in development of irrigated area, the overall performance of the irrigation sector has not been very commendable in India. While the water use efficiency is very low (ranging from 30-40 per cent under surface method of irrigation) in India as compared to other countries (Rosegrant, 1997), the financial performance of the state managed irrigation sector has been deteriorating over the years. The per cent of recovery rate (ratio of irrigation revenue to operation and maintenance cost) of irrigation and multipurpose river valley projects, which are completely managed by the state agency, has declined from 36.4 percent in the 1976-77 to as low as 5.7 per cent in 1999-2000[2]. There are many reasons for the dreadful performance of the irrigation sector. Significant increase in operation and maintenance (O&M) expenditures of the irrigation sector and unrevised irrigation water rates charged from farmers are often cited as the main reasons for the poor financial performance of the publicly funded irrigation sector (GOI, 1992; Gulati, et al., 1994; Vaidyanathan, 1999). Many argue that the poor water use efficiency, which exists across different regions in India, is the result of low water price fixed for different crops (Vaidyanathan, 1999; World Bank, 2005). Poor participation of farmers in irrigation management related activities is also cited as one of the major reasons for the deteriorating performance of irrigation sector in India. Researchers working on irrigation related issues in different countries suggest that the poor performance of irrigation sector can be improved substantially by extensively involving the users (farmers) in the management activities of the irrigation sector (Brewar, et al., 1999).

Role of Water Users' Association in Irrigation Management

Involving farmers in water management activities is not a new idea as far as India is concerned. In the name of *kudimaramath*, the users (farmers) were involved in tank repairs, management works, distilling of tanks, removal of weeds, distribution of water and collecting revenue even during the 17th and 18th century, much before the British rule (Reddy, 1996, Maloney and Raju, 1994). Unfortunately, due to various reasons, the users' participation in water management has declined drastically over the years. Now, partly because of pressure from donor agencies such as the World Bank and others, the users' participation in irrigation water has become a widespread strategy in all countries in Asia including India at least from the mid-eighties (Vermillion, 1992 and 1997; Easter, 2000). When one addresses the users' participation in irrigation management, the following questions automatically crop-up which are essential to understand before getting into further details of participatory irrigation management:

- Why do we need to encourage users' participation in irrigation management?
- Can the users' participation solve the problems that are currently faced by the Indian irrigation sector?
- How far users' participation will help to improve the financial position of publicly funded irrigation sector?
- Can users' group accomplish their duties without the help of government?
- Given the large size of India's irrigation sector, is it worthwhile to encourage the users' participation in irrigation sector?
- Given the inequality in land distribution, can the users' group perform their duties without any pressure from large size farmers?
- Can the users' group perform duties without any proper legal backup?

The ultimate aim of the decentralized water management with predominant users' participation is to improve the overall performance of irrigation sector. In India, much emphasis was given for users' participation only after the announcement of the first National Water Policy: 1987, where gradual involvement of farmers in system management was advocated. The National Water Policy states that "efforts should be made to involve farmers progressively in various aspects of management of irrigation systems, particularly in water distribution and collection of water rates" (MOWR, 1987). The second National Water Policy of India: 2002 also reiterates the importance of farmers' participation in irrigation management activities (MOWR, 2002). It emphasises that "Management of the water resources for diverse uses should incorporate a participatory approach; by involving not only the various governmental agencies but also the users and other stakeholders, in an effective and decisive manner, in various aspects of planning, design, development and management of the water resources schemes. Necessary legal and institutional changes should be made at various levels for the purpose, duly ensuring appropriate role for women. Water Users' Associations and the local bodies such as municipalities and gram panchayats should particularly be involved in the operation, maintenance and management of water infrastructures/facilities at appropriate levels progressively, with a view to eventually transfer the management of such facilities to the users groups/local bodies".

Farmers Participation in Water Managment

Why has the government decided to involve farmers in water management activities? Apart from aid agency pressures, there could be two main reasons for this. First, despite the introduction of Command Area Development Programme (CADP) during 1974-75 (fifth five-year plan) for increasing the water use efficiency, it could not make any big change in the existing water use efficiency as of today (Brewer, et al., 1999)[3]. Second, the financial performance of irrigation sector of India has deteriorated over the years and the present collection of revenue from irrigation sector is not even enough to manage the operation and maintenance cost of irrigation sector (GOI, 1992; CWC, 2004). Under these circumstances, the experiences of the countries where farmers were allowed to involve in system management showed very encouraging results in both water management and financial recovery (see, Tables 1.1, 1.2 and 1.3). This has helped the policy makers to encourage farmers' participation in system management in a big way.

Table 1.1: Impacts of Irrigation Management Transfer on Operation and Maintenance

Study, Country, type of irrigation	*Operations*	*Maintenance*
Oorthuizen and Kloezen (1995), Philippines, SI	Water distribution became less equitable.	Worsened
Wijayaratna & Vermillion (1994) Philippines, SI	Improved water distribution equity; expansion of dry-season irrigated area.	N.A.
Bagadion (1994) Philippines, LI	No increasing trend in service area.	N.A.

Svendsen (1992) Philippines, SI, LI	Equity of water distribution improved and 7% expansion of benefited area in the dry season.	N.A.
Johnson & Reiss (1993) Indonesia, LI	N.A.	Deterioration of pump sets accelerated.
Nguyen & Luong (1994) Vietnam, LI	Water consumption per hectare dropped 36%. Area irrigated increased 71%	N.A.
Johnson et al. (1995) China, SI	Reduction in water duty from 11,000 m^3 to 4, 500 m^3 .	N.A.
IIMI & BAU (1996) Bangladesh, LI	Declining numbers of farmers reporting adequate and timely water delivery.	Higher breakdown rates in smaller pumps; spare parts and repair easier.
Olin (1994) Nepal, LI	Drop in water consumption by 50%.	N.A.
Rana et al. (1994), Nepal, SI	Irrigation discharge increased fourfold.	N.A.
Mishra & Molden (1996) Nepal, SI	Inflow increased from 2.2 m^3/s to 7.9 m^3/s (26%-93% of design capacity)	N.A.
Kloezen (1996) Sri Lanka, SI	Quality of water distribution did not change	N.A.
Uphoff (1992) Sri lanka, SI	Improved equity of water distribution.	Maintains activity and investment increased.
Pant (1994) India, LI	Reduced average irrigation time.	Maintenance work increased.
Srivastava & Brewer (1994) India, SI	Improved equity; 27% more water tail end; 20% increase in irrigated area in the dry season.	More maintenance activity.
Rao (1994), India, SI	Improved equity of water distribution	N.A.

Kalro and Naik 1995 India, SI, LI	Adequacy and reliability of water distribution improved, as reported qualitatively.	N.A.
Azziz (1994) Egypt, SI	Reduced irrigation time; better water adequacy.	N.A.
Samad & Dingle (1995) Sudan, LI	Timeless and water adequacy worse in schemes turned over.	N.A.
DSI, EDI, & IIMI (1996) Turkey, SI, LI	In first year, area served increased 20-40%.	N.A.
Wester, During, & Oorthuizen (1995) Senegal, LI	Expansion of areas irrigated.	Deterioration of pump sets accelerated.
Maurya (1993); Musa (1994) Nigeria, SI	Improved equity; 12% more water reached middle and tail reaches.	Increased maintenance activity.
Yap-Salinas (1994) Dominican Republic, SI	Delivery efficiency improved 25-30%.	N.A.
Vermillion & Garces-Restrepo (1996) Colombia, SI	More responsive operations. Water adequacy satisfactory. 40-45% of farmers say operations improved. Temporary inefficiencies after IMT. Expansion of area irrigated.	Good maintenance; 92-98% of farmers report quality of maintenance has not changed.
Johnson (1996), Mexico, SI	No change in water delivered per hectare or in area irrigated	N.A
Svendsen & Vermillion (1994) USA, SI	More responsive operation. Efficiency did not change. Equity improved slightly.	Good but slight trend detected.

Notes: SI- surface irrigation, LI- lift irrigation, N.A. - not available.
Source: Vermillion, (1997).

Table 1.2: Impacts of Irrigation Management Transfer on Agricultural and Economic Productivity

Study, country, type of irrigation	*Agricultural Productivity*	*Economic productivity*
Wijayaratna & Vermillion (1994) Philippines, SI	Increase in cropping intensity	N.A.
Svendsen (1992) Philipppines, SI, LI	Rice yield increased by 4% to 4 t/ha in both wet and dry seasons.	N.A.
Nguyen & Luong (1994) Vietnam, LI	Cropping intensity increased from 170% to 250%. 14% increase in area cropped. Yield increased 13%.	Annual incremental benefits increased by US $193/ha or by $182/ha net of increased O & M cost
Johnson et al. (1995) China, SI	Grains yield increased modestly.	Cases of both increase and decrease in net income.
IIMI & BAU (1996) Bangladesh, LI	Slight increase in cropping intensities. Mixed results for yields.	Small farmers (<1 ha) becoming a growing share of pump owners and of expanding irrigated area (58% to 63%, 1989-94).
Mishra & Molden (1996) Nepal, SI	Rice yields increased from 2.2 t/ha to 3.4 t/ha. Wheat yields increased from 1.6 t/ha to 2.4 t/ha.	N.A.
Kloezen (1996) Sri Lanka, SI	Cropping intensities increased from 138% to 200%	Gross annual value of output between US$944/ha and $1,136/ha after IMT.
Pant (1994) India, LI	Cropping intensity increased from 143% to 162%. Yields increased 10%.	N.A.

Kalro and Naik (1995) India, SI, LI	Increases in cropping intensities and crop diversification. No change in yields.	N.A.
Azziz (1994) Egypt, SI	10-16% increase in main crop yields.	Increase in farm incomes by US $60/ ha
Samad & Dingle (1995) Sudan, LI	High yields per unit of water in parastatal schemes (17 kg/100 m^3) versus turned-over schemes (11 kg/100 m^3)	Gross margin three higher in parastatal than in turned-over schemes. Productivity of land and water higher in parastatal than in turned-over schemes.
Maurya (1993); Musa (1994) Nigeria, SI	Increase in dry-season cropped area by 80%.	N.A.
Wester, During, & Oorthuizen (1995) Senegal, LI	Cropping intensity rising and falling in different locations.	Cost of irrigation rice production increased 78%.
Vermillion & Garces-Restrepo (1996) Colombia, SI	Rice yields of 6.5 t/ ha, sustained after IMT. Cultivated area continued to expand. More crop diversification.	Net farm income rose 23%. Economic return to irrigation was US $11-$12/100 m^3 water. Gross value of output increased 400%, 1983-91.
Graces-Restrepo & Vermillion (1994), Colombia, SI, LI	N.A.	Cost of water relative to cost of rice production increased from 2.0% to 2.4%.
Johnson (1996), Mexico, SI	No change in cropping intensity or yields.	Annual economic returns (US $ 1, 500-$1,900/ha) remained same or declined.

Svendsen & Vermillion (1994) USA, SI	Shift to les water-intensive crops but more due to changing water application technology and markets.	Average farm incomes rose 15% due to reduction in water cost.

Notes: SI- surface irrigation, LI- lift irrigation, n. a. - not available.
Source: Vermillion, (1997).

Table 1.3: Impacts of Irrigation Management Transfer on Irrigation Finance

Study, Country, irrigation type	*Impacts*
Oorthuizen and Kloezen (1995), Philippines, SI	Reduced cost to farmers; 75% drop in budget. Fee collection rates rose from 20% to 81%.
Wijayaratna & Vermillion (1994) Philippines, SI	Revenue from water charges increased from 24% in 1979 to 60% in 1990. Reduction in agency field staff. Diversification of revenue sources.
Bagadion (1994) Philippines, LI	Budget losses converted to surpluses. Fee collection rate rose from 27% to 60%.
Svendsen (1992) Philipppines, SI, LI	Decrease in frequency of deficit budgets; increase in revenue from water charges and other income. 29% drop in the operating expenditures. Decline in staff from 13% to 75%. Government subsidy dropped from 25 million in 1976 to zero in 1982.
Johnson & Reiss (1993) Indonesia, LI	Cost of water pumped increased five to seven times.
Johnson et al (1995) China, SI	Per hectare cost of water to farmer rose 2.5 times. Growing importance of sideline revenue enterprises after reform.
IIMI & BAU (1996) Bangladesh, LI	O & M costs remained similar after privatization, though loss of subsidies meant increased costs to farmers. Diesel deep tube wells not financially viable without subsidy.

Olin (1994) Nepal, LI	Cost of water decreased 40-50%.
Mishra & Molden (1996) Nepal, SI	Cash and labor value raised from farmers increased to US$6.77/ha, and 77% of farmers paid water charges.
Kloezen (1996) Sri Lanka, SI	Government subsidies for O & M continued. Farmer organizations invested mainly in input provisions and marketing, not in O & M. Annual government costs decreased 33%. Diversification of revenue sources.
Pant (1994) India, LI	50% reduction in the cost of water. Budget deficits converted to surplus.
Shah, et al. (1994) India, LI	50% reduction in the cost of water.
Kalro and Naik (1995) India, SI, LI	Increased costs and time required for farmers. Improved rate of recovery for water charges. No decline in government expenditures for O & M.
Azziz (1994) Egypt, SI	Dramatic decline in maintenance costs. Per hectare pumping costs declined from US$68-$79 to $45-$50 after rehabilitation and IMT.
DSI, EDI, & IIMI (1996) Turkey, SI, LI	Increase in average water fee collection rate from 38% to 72% in the first year after IMT.
Maurya (1993) Nigeria, SI	Water fee collection rates rose from 50% to 90% after IMT.
Yap-Salinas (1994) Dominican Republic, SI	Water fees increased 1,500% in 8 years. Fee collection rates increased from 12% to 80%.
Vermillion & Garces-Restrepo (1996) Colombia, SI	44% average decline in total staff. Farmers emphasis on cost-cutting. No long-term major change in cost of irrigation. Cost of water relative to production fell 27%. Diversification of revenue sources, from 10% to 20% of revenue other than fees. Budget deficits converted to surpluses.
Garces-Restrepo & Vermillion (1994) Colombia, SI, LI	Declining trend in fee levels. Reduction in staff by 38%. Shift from deficit to surplus budgets in all study cases.

Johnson 1996; Gorriz, Subramanian, & Simas (1995) Mexico, SI	45-180% increase in water charges. Increase in fee collection rates from 15%, originally, to 80% to 100%. Shortfall in financing declined nationally from US$66 million to $41 million annually. Local self-reliance increased from 43% to 78%.
Svendsen & Vermillion (1994) USA, SI	Decrease in government staff by 86%. Farmers emphasis on cost-cutting. Volumetric charges reduced by 16%. Diversification of revenue sources. Water charge was 80% of revenue before and 67% after IMT. Farmers raised capital replacement fund.
Farley (1994) New Zealand, SI	Farmer emphasis on cost-cutting. Average operational costs declined 66%. After IMT, water charges were a quarter to a half of the pre-IMT level.

Notes: SI- surface irrigation, LI- lift irrigation.
Source: Vermillion, (1997).

Positive Impacts

In so far as the results are concerned, it has been clearly established that users-managed systems outperformed the systems which are managed by the irrigation agency all over the world including India (Easter, 2000; Vermillion, 1997). The most often found positive impacts of users-managed systems are:

(a) Reduction in the cost of irrigation to farmers and government;

(b) Enhanced financial self reliance of irrigation schemes;

(c) Expansion of irrigation;

(d) Reduction in the amount of water delivered per hectare;

(e) Significant increase in cropping intensity and yield of crops.

Negative Impacts

Undeniably, the transfer of irrigation system from government agency to farmers (users) group has also made some negative impacts in certain places as well. The often cited negative impacts are increased costs to farmers, failing financial viability of lift schemes and deteriorating infrastructure (Vermillion, 1997). Though majority studies show positive impacts due to users managed systems, most of the studies have not applied proper methodology for assessing the performance. It is rightly pointed out by Vermillion (1997) "........much of the so-called evidence about impacts is based upon qualitative reports, without before-and-after or with-and-without comparisons, without independent measurements with quantitative data, without systematic sampling of farmers, and a failure to distinguish possible alternative causes of the observed 'impacts' of transfer" (p. 29).

Overall Assessment

Though the transfer of system management has increased area under irrigation and productivity gains, the coverage of area by water users association is very meagre when compared to the total irrigated area so far. Vaidyanathan committee on pricing of irrigation water observes that "the area covered by these initiatives is very small, less than 1 per cent of area irrigated at present. The general consensus among knowledgeable people is that they have been fitful and have not really made much of an impact. For the most part the outlet and canal committees are there only in name; they are not consulted on substantive issues; nor are department officers required to follow their advice. There is also considerable reluctance, if not oppositions, from the operational staff of irrigation departments to involving users in management; and even users themselves tend to be apathetic to the idea" (GOI, 1992, pp.126-127).

Water users associations are functioning mostly at the tertiary level, which cannot accomplish their duties in improving supply and management of irrigation as the supply of water is controlled by the irrigation agency. Users' organisations that exist in small irrigation systems namely tank, etc., could perform their duty relatively better than WUAs that are established in large irrigation network namely canal. There are two reasons for this. First, the control of government agency is very meagre under small irrigation systems like tanks. Second, users group can clearly understand the demand and supply position of water and make decisions accordingly.

As regards WUAs in larger irrigation systems, the overall results are not encouraging so far. Turnover of irrigation systems has been slow in most of the large irrigation projects in different parts of the world including India. There are reports that staffs who are managing irrigation systems see these Association as potential threat to their jobs (Vermillion, 1997; Easter, 2000). Therefore, the wholehearted involvement in establishing and supporting the Association by the agency staff may not be very high.

Moreover, unlike other South Asian countries, irrigation systems are very large in India which is practically not possible to manage efficiently by WUAs without adequate support from the irrigation agency. In order to encourage the farmers' attachment with these Association, it is essential to demonstrate the benefits in delivering water supply at required quantity and time. Importantly, WUAs should be able to reduce the cost of water over time in order to show the advantages of these Associations to the farmers (Vermillion, 1997).

WUAs have to be legally established in order to increase their responsibility and decisions related to water management.[4] It is always difficult for WUAs to provide better water supply and other services to their members without legal standing (Easter, 2000). With an improved service, WUAs can convince the farmers to pay the charges for the water that they use. As the Committee on Pricing of Irrigation Water (GOI, 1992) rightly observes that "An essential pre-condition is to convince users that they will benefit from such group activity by getting more water, more assured supplies according to a pre-specified schedule (or according to the needs of the crops), greater flexibility in the use of water, or some combinations of these. Improvement in any of these dimensions will almost certainly increase productivity and therefore induce farmers to take the idea of users' groups more seriously". Continued support from government agencies is essential even after transforming the systems management to users group in order to sustain the participation of farmers. Importantly farmers at any level should not be allowed to think that the transfer of irrigation management is adopted in order to reduce the financial burden of the government.

Studies on PIM in Maharashtra State

Since the focus of the work is on Maharashtra State, let us briefly look at the findings of the existing studies on participatory irrigation management. As in the case of other countries, where participatory irrigation management in operation, quite a few studies have been carried out to assess the impact of PIM on various parameters in Maharashtra State as well. The experiment of participatory irrigation management started during 1987 by farmers, sugar cooperatives and government officers. But, none of these experiments could sustain for a long time due to various reasons (see, Naik, et al., 2002).

A well known experiment was carried out by the Centre for Applied Systems Analysis in Development (CASAD) in 1987 by establishing WUAs in Mula Right Bank Canal in Ahmednagar district. In order to document the process of formation of WUAs and the benefits, Lele and Patil (1994) studied the four years experience of organising and operationalizing WUAs. The study shows that the member farmers were able to get many tangible and intangible benefits because of the formation of WUAs. Farmers were able save their journey to get water, payments were simplified, water was made available at the designated time, which reduced the uncertainty of waiting and increased the recovery rate of water charges significantly. Importantly, the authors report that "There were hardly any complaints from farmers regarding water. Every person who demanded water got it in time, unlike earlier" (Lele and Patil, 1994, p. 119).

Impact of Irrigation Management Transfer

Since Maharashtra has had over 76,000 ha area under the control of 228 WUAs even in the late nineties, Pant (1999) made an attempt to assess the impact of irrigation management transfer in Maharashtra using the data available with the various government departments dealing with the irrigation sector. The analysis of Pant (1999) shows that the establishment of large WUAs in the state has helped to increase the irrigated area as well as water use efficiency. Though WUAs charges much higher water rates from the member water users, it could considerably increase the recovery rate of water charges and thereby bringing more revenue to government. However, Pant (1999) observes that "Given the extent to which IMT is being implemented in Maharashtra, it is frustrating to find that there is very little information available about the impact of IMT programme on the aspects detailed in the preceding. This is more so when there is a full fledged directorate in Maharashtra to monitor the IMT programme which regularly collects information on cropwise irrigated area, sanctioned and used water, water charges collection, etc. (Pant, 1999. p. A20).

Unlike Pant (1999), using the field data collected from two irrigation societies namely Shri Datta Society in the Mula Scheme and Laxmi Narasimha Society in the Bhima Scheme, Naik et al., (2002) studied the impact of irrigation management transfer in Maharashtra on various parameters. This study finds significant impact on the yield of crops, government finances, maintenances of transferred canals, distribution of water due to the establishment of WUAs. However, this study notes a considerable variation in the impact parameters between the two selected WUAs for the study, mainly because of variation in the supply of water. It suggests that the impact of WUAs on different parameters tends to vary depending upon the factors that govern the users association.

Impact of Participatory Irrigation Management

A study carried out in the Vidarbha region of Maharashtra, which is one of the less irrigated regions in the state, also shows substantial benefits from participatory irrigation management programme. Belsare (2001), while studying the impact of PIM in Katepurna irrigation project located in the Vidarbha region, finds that the impact of PIM during the year 1998-99 to 2000-01 was substantial in increasing the area under irrigation. The area under irrigation in the project increased from 2027 ha to 3646 ha during the year, with yearly water saving of about 7.71 mm^3. Belsare (2001) underlines that for the first time in the history of Katepurna project, the project achieved irrigation equal to its present potential with 86 per cent of live storage, because of extensive participation of farmers in water management activities.

In Maharashtra, WUAs are operating not only in canal command areas but also in the areas where lift irrigation is used for cultivation. In order to mitigate the drought situation and to reduce poverty levels in rural areas, there are several new water institutions established in different regions in India with the support of NGOs and other voluntary organizations especially since early 1980 onwards.

Pani Panchayat Institution

One of the new water institutions emerged during the seventies in the rainfed areas of Maharashtra State is Pani Panchayat (water council).[5] This institution is not promoted by the State agencies, but established by a Trust (Gram Gourav Prathishtan) with the voluntary support from users' group. Deshpande and Reddy (1990) studied the overall functioning as well as the impact of Pani Panchayat on different parameters. This study shows that the establishment of Pani Panchayat has increased the irrigated area and also changed the cropping pattern from low value crops to high value crops in this area and the yield rate was also improved substantially (Deshpande and Reddy, 1990). However, owing to breakdown of Pani Panchayat Institutions and other technical reasons, presently most of the Pani Panchayat Institutions are not performing well. Recent studies by Thakur and Pattnaik (2002) and Keremane, et al., (2006) have highlighted the reasons for the decline of Pani Panchayat Institutions in a detailed manner.

General Status of PIM in Maharashtra

Maharashtra State has long history with regard to Participatory Irrigation Management (PIM). Farmers managed irrigation system such as

"Phad"[6] system and "Malgujari"[7] tank irrigation system are very popular over centuries in Maharashtra. However, the new form of PIM started picking up only after the mid-eighties, precisely after the announcement of the first National Water Policy: 1987. Some of the non-governmental organisations have initiated the formation of WUAs in certain areas in the state, which produced exemplary results in terms of expansion of irrigated area, recovery rate, desirable cropping pattern, etc., (see, Lele and Patil, 1994). These encouraging results in a way prompted farmers and NGOs to increase the participation of farmers in irrigation water management activities in the State. In Maharashtra, however, significant progress has taken place since 1992 as the irrigation department has been encouraging the farmers to form WUAs by explaining its advantages.

DIRD Activities

Directorate of Irrigation Research and Development (DIRD) of Water Resources Department, Government of Maharashtra, Pune actively engages with the formation of these Associations. It involves in four important activities pertaining to Them:

(1) Collection of information required to establish WUAs from the irrigation management division at various stages such as proposal, registration, agreement, joint inspection, handing over, etc.
(2) Arrange seminars, workshops and deals with the problems of WUAs or irrigation officers regarding procedures and policy matters.
(3) Preparation of bimonthly status report on the activities. The DIRD has prepared a clause and published a guidebook for WUAs as well.[8]

In terms of formation of WUAs, the state has made rapid progress over the last one decade. In order to provide legal recognition to the water users association, the state has enacted an Act[9] which is known as Maharashtra Management of Irrigation System by Farmers Act, 2005.[10]

Table 1.4 presents the details of number of WUAs established and the area controlled by the WUAs for the state as a whole. It is evident from the table that there has been a significant progress in the number of WUAs functioning/registered/proposed since September 1996. While the number of WUAs currently fully functioning increased from 100 in September 1996 to 774 in July 2005, the area controlled by these WUAs increased from about 43.68 thousand hectares to 2.51 lakh hectares during this period. That is, the area covered by the fully functioning WUAs has increased nearly by six times during this period, which is appreciable. If we include all the WUAs which are currently signed/registered/proposed

Table 1.4: Present Status of Water Users' Association in Maharashtra State

Particulars	*As of September 1996*		*As of December 2003*		*As of July 2005*	
	Nos.	*CCA (Ha.)*	*Nos.*	*CCA (Ha.)*	*Nos.*	*C CA (Ha.)*
Water Users' Association functioning	100	43684	533	158923	774	2 50521
Agreement Signed, yet to Hand Over	34	9894	129	46367	426	1 51936
Registered yet to Sign Agreement	180	60372	963	347399	1201	4 33410
WUAs Proposed	143	55211	847	244898	1650	7 04948
Total	457	169105	2472	797587	4051	15 40815

Sources: Naik, *et al.*, (2002); GOM (various years) and *www. mahagovid.org.*

Table 1.5: Region-wise Status of Water Users' Association as of July 2005

Sr. No.	Region	*Water Users' Association functioning*		*Agreement Signed, yet to Hand Over*		*Registered yet to Sign Agreement*		*Proposed*		*Total*	
		Nos.	*CCA (Ha.)*	*Nos.*	*CCA (Ha.)*	*Nos.*	*CCA (Ha.)*	*Nos.*	*CCA (Ha.)*	*Nos.*	*CCA (Ha.)*
	(A) Major and Medium Projects										
1.	Konkan	0 (0.0)	0 (0.0)	0 (0.0)	0 (0.0)	4 (0.3)	600 (0.1)	12 (0.7)	4680 (0.7)	16 (0.4)	5280 (0.3)
2.	Nagpur	36 (4.7)	12790 (5.1)	64 (15.0)	24990 (16.4)	176 (14.7)	80033 (18.5)	501 (30.4)	258176 (36.6)	777 (19.2)	375989 (24.4)
3.	Amravati	112 (14.5)	29735 (11.9)	69 (16.2)	20753 (13.7)	177 (14.7)	48421 (11.2)	411 (24.9)	186210 (26.4)	769 (19.0)	285119 (18.5)
4.	North Maharashtra	176 (22.7)	61009 (24.4)	140 (32.9)	40602 (26.7)	158 (13.2)	52929 (12.2)	66 (4.0)	19579 (2.8)	540 (13.3)	174119 (11.3)
5.	Pune	234 (30.2)	60077 (24.0)	20 (4.7)	6638 (4.4)	204 (17.0)	54934 (12.7)	305 (18.5)	107763 (15.3)	763 (18.8)	229412 (14.9)
6.	Marathawada	106 (13.7)	46415 (18.5)	52 (12.2)	20745 (13.7)	251 (20.9)	126367 (29.2)	58 (3.5)	34285 (4.9)	467 (11.5)	227812 (14.8)
	Total	664 (85.8)	210026 (83.8)	345 (81.0)	113728 (74.9)	970 (80.8)	363284 (83.8)	1353 (82.5)	610693 (86.6)	3332 (82.3)	1297731 (84.2)

	(B) Minor Irrigation Projects										
1.	Konkan	8 (1.0)	958 (0.4)	5 (1.2)	604 (0.4)	14 (1.2)	2263 (0.5)	47 (2.8)	8203 (1.2)	74 (1.8)	12028 (0.8)
2.	Nagpur	1 (0.1)	165 (0.1)	25 (5.9)	14972 (9.9)	25 (2.1)	8274 (1.9)	53 (3.2)	19856 (2.8)	104 (2.6)	43267 (2.8)
3.	Amravati	66 (8.5)	21626 (8.6)	17 (4.0)	5795 (3.8)	125 (10.4)	37894 (8.7)	158 (9.6)	58758 (8.3)	366 (9.0)	124073 (8.1)
4.	North Maharashtra	14 (1.8)	5483 (2.2)	2 (0.5)	670 (0.4)	37 (3.1)	11157 (2.6)	7 (0.4)	1881 (0.3)	60 (1.5)	19191 (1.2)
5.	Pune	2 (0.3)	177 (0.1)	0 (0.0)	0 (0.0)	4 (0.3)	1268 (0.3)	19 (1.2)	3124 (0.4)	25 (0.6)	4569 (0.3)
6.	Marathawada	19 (2.5)	12086 (4.8)	32 (7.5)	16167 (10.6)	26 (2.2)	9270 (2.1)	13 (0.8)	2433 (0.3)	90 (2.2)	39956 (2.6)
	Total	110 (14.2)	40495 (16.2)	81 (19.0)	38208 (25.1)	231 (19.2)	70126 (16.2)	297 (18.0)	94255 (13.4)	719 (17.7)	243084 (15.8)
	Total (A+B)	774 (100.0)	250521 100.0)	426 (100.0)	151936 100.0)	1201 (100.0)	433410 (100.0)	1650 (100.0)	704948 (100.0)	4051 (100.0)	1540815 (100.0)

Note: Figures in brackets are percentage to total.
Source: *http://www.mahagovid.org/default_en.htm*

Table 1.6: Region-wise Share of Area under WUAs to Surface Irrigated Area

Irrigation circles	*Area under fully functioning WUAs (ha)*	*Area to be brought under WUAs (ha)*	*Total Area* (ha)*	*Surface Irrigated Area (ha) TE 2001-02*	*(2)/(5)%*	*(3)/(5)%*	*(4)/(5)%*
(1)	(2)	(3)	(4)	(5)	(6)	(7)	(8)
Konkan	958	16350	17308	25433	3.77	64.29	68.05
Nagpur	12955	406301	419256	235167	5.51	172.77	178.28
Amravati	51361	357831	409192	47733	107.60	749.65	857.25
North Maharashtra	66492	126818	193310	197900	33.60	64.08	97.68
Pune	60254	173727	233981	314800	19.14	55.19	74.33
Marathawada	58501	209267	267768	229067	25.54	91.36	116.90
Total	250521	1290294	1540815	1050100	23.86	122.87	146.73

Note: * Total of area proposed, registered to sign agreement and agreement signed, yet to hand over.

Sources: http://www.vmahagovid.org/default_en.htm and GOM (various years).

for the analysis, the achievement made by the State government would go up significantly further. The total number of WUAs increased from 457 in September 1996 to 4051 as of July 2005 and the area covered by these WUAs increased from about 1.69 lakh hectares to about 15.41 lakh hectares in July 2005.

Is the spread of WUAs same across the regions in the State? Data presented in Table 1.5 shows that the spread of WUAs is not the same across different regions in the State. The irrigation department of Maharashtra classifies the whole of Maharashtra under six irrigation circles namely Konkan, Nagpur, Amarawati, North Maharashtra, Pune and Marathwada. Among the six irrigation circles, Pune circle has more number of fully functioning WUAs as compared to other circles. For instance, out of the 774 fully functioning WUAs in the State, Pune circle accounts for about 30 per cent of WUAs followed by North Maharashtra (24 per cent), Amarawati (23 per cent) and others. However, in terms of area currently controlled by WUAs, North Maharashtra has relatively larger share in comparison to Pune and other circles. This could be possibly because of larger size of WUAs in North Maharashtra as compared to Pune circle.

Since WUAs are primarily established in canal command areas, we expect that there could some positive relationship between the total area under surface irrigation and the area managed by WUAs in each region. However, the data presented in Table 1.6 shows somewhat different relationship between the two. Be that as it may, the available data from irrigation department clearly shows that there is a significant progress in the formation of WUAs in the State. It would be useful to study how these WUAs are functioning at the field level in carrying out specified tasks and objectives. In this study, therefore, an attempt is made to find out the overall functioning of WUAs as well as its impact on different parameters using the data collected from 150 sample farmers registered with 10 different types of WUAs.

Objectives of the Study

1. To study what extent WUAs have contributed towards regular water supply, efficiency in water use, collection of water charges, and operation and maintenance of the water delivery systems.
2. To study the impact of WUAs on cropping pattern, input use and improvement in productivity of crops as well as in the socio-economic well being of the beneficiaries.
3. To study what extent the devolution of power to WUAs has taken place in the State.

4. To identify the constraints in the effective implementation of PIM and ways to remove the same.
5. To suggest policy recommendations for the effective implementation of PIM.

Framework

What is the ideal condition under which the local level collective action would be successful in managing the common property resources? This is one of the issues of the common property resource management, which has attracted a large number of researchers around the world in recent years (see, Hardin, 1982; Bardhan, 1993; Wade, 1987; 1995; Seabright, 1993, Baland and Platteau, 1997; Bardhan and Dayton-Johnson, 2002). Many studies are available on this area, but the results of the studies are not the same. While one group of researchers argue that the cooperative behaviour is difficult to sustain in an environment which has large heterogeneous agents (Seabright, 1993; Kanbur, 1992), the other group of researchers argue that heterogeneity does not necessarily matter in bringing collective participation in the commons (Olson, 1965; Wade, 1994; Baland and Platteau, 1999). In fact, Olson (1965) hypothesised that groups where considerable heterogeneity exists may be more successful if those with the most economic interests and power initiate collective action.

Specific studies have also been carried out relating to the performance of community based irrigation systems and group heterogeneity using data from different countries (Tang, 1991; Lam, 1998; Bardhan, 2000; Bardhan and Dayton-Johnson, 2002). However, except the study of Bardhan (2000) which looked at the cooperative behaviour of 48 irrigation communities in South India, not many studies are available focusing on this issue in Indian context. In the present study, therefore, an attempt is made to find out the heterogeneity impact of WUAs on different developmental parameters including the participation of the farmers in water management and other related activities. As mentioned earlier, quite a few studies have looked at the relationship between group members' heterogeneity and collective action in the commons. In the present study, besides looking at the heterogeneity of the members of WUAs, we also make an attempt to see as to how the heterogeneity in physical characteristics such as location of the Associations and source of water used by WUAs affect the participation of farmers in water management activities and other impact parameters.

In Maharashtra State, while most of WUAs are formed with the

Table 1.7: Trends in Irrigated Area - Maharashtra and India

(Area in mha)

Period	SIA		WIA		NIA		GIA	
	Maharashtra	*India*	*Maharashtra*	*India*	*Maharashtra*	*India*	*Maharashtra*	*India*
TE 1962-63	0.48 (44.04)	17.64 (69.29)	0.61 (55.96)	7.43 (29.64)	1.09 (100)	25.07 (100)	1.24	28.63
TE 1972-73	0.58 (43.61)	19.12 (60.70)	0.75 (56.39)	12.38 (39.28)	1.33 (100)	31.49 (100)	1.55	38.56
TE 1982-83	0.79 (41.36)	21.38 (53.49)	1.12 (58.64)	18.59 (46.51)	1.91 (100)	39.97 (100)	2.44	51.01
TE 1992-93	0.97 (36.06)	23.18 (46.93)	1.71 (63.94)	25.88 (53.08)	2.69 (100)	49.39 (100)	3.28	65.22
TE 2000-01	1.05 (35.35)	21.40 (39.14)	1.92 (64.65)	33.28 (60.86)	2.97 (100)	54.68 (100)	3.66	75.14
TE 2003-04	1.04 (35.13)	20.75* (38.03)	1.96 (64.87)	33.81* (61.97)	2.96 (100)	54.55* (100)	3.66	75.35*

Notes: * - relates to TE 2002-03; SIA – surface irrigated area; WIA – well irrigated area;
Figures in brackets are percentages to net irrigated area (NIA); GIA – Gross Irrigated Area.
Sources: GOI (various issues); GOM (various issues); FAI (various years); CMIE (2006).

Table 1.8: Division-wise Trends in Area under Irrigation in Maharashtra (Area in '000 ha)

Division	*SIA*		*WIA*		*NIA*		*GIA*	
	TE 1962-63	*TE 2001-02*	*TE 1962-63*	*TE 2001-02*	*TE 1962-63*	*TE 2001-02*	*TE 1962-63*	*TE 2001-02*
Konkan	10.97 (2.28)	25.43 (2.42)	9.77 (1.59)	27.40 (1.43)	20.77 (1.90)	52.83 (1.78)	21.37 (1.72)	63.73 (1.74)
Nasik	29.30 (6.10)	108.13 (10.30)	84.33 (13.77)	302.43 (15.76)	113.77 (10.41)	410.57 (13.83)	147.33 (11.84)	511.20 (13.97)
Pune	121.47 (25.28)	246.23 (23.45)	233.30 (38.09)	479.60 (25.00)	354.90 (32.47)	725.83 (24.45)	413.27 (33.22)	886.00 (24.22)
Kolhapur	75.80 (15.78)	158.33 (15.08)	87.17 (14.23)	249.50 (13.00)	162.97 (14.91)	407.83 (13.74)	194.37 (15.63)	495.90 (13.55)
Aurangabad	5.93 (1.23)	139.77 (13.31)	85.67 (13.98)	310.17 (16.17)	91.63 (8.38)	449.93 (15.16)	107.47 (8.64)	554.03 (15.14)
Latur	12.53 (2.61)	89.30 (8.50)	63.13 (10.31)	184.47 (9.62)	67.47 (6.17)	273.77 (9.22)	75.57 (6.08)	348.33 (9.52)
Amaravati	0.53 (0.11)	47.73 (4.55)	25.97 (4.24)	129.00 (6.72)	26.57 (2.43)	176.73 (5.95)	26.63 (2.14)	225.03 (6.15)
Nagpur	231.87 (48.26)	235.17 (22.39)	23.27 (3.80)	235.93 (12.30)	255.23 (23.35)	471.10 (15.87)	257.50 (20.70)	574.60 (15.70)
State	480.50 (100.00)	1050.10 (100.00)	612.57 (100.00)	1918.50 (100.00)	1093.10 (100.00)	2968.60 (100.00)	1243.87 (100.00)	3658.83 (100.00)

Note: Figures in bracket are percentage share to state total. *Sources*: GOI (various issues), GOM (various issues).

support of State agencies especially after mid-nineties, Pani Panchayat institutions are established with the initiatives of some non-governmental organisations. Unlike majority of the Indian States, WUAs have been established in Maharashtra State not only in canal command areas (at various locations namely head, middle and tail end), but also in lift irrigated areas. Maharashtra State has very rich experience of lift irrigation cooperative societies, which have been functioning at different locations of water sources namely river, well and backwater of the reservoir/ dam. The present study covers WUAs with different characteristics and therefore, it is expected to bring out whether heterogeneity in physical characteristics matters in determining the functioning of WUAs.

Information Collection and Technique[11]

The study has been carried out in Maharashtra State, which has the largest irrigation sector in India both in terms of investment and number of irrigation projects (see, Deshpande and Narayanamoorthy, 2001). Despite this, the coverage of irrigation to cropped area in the State is only around 17 per cent as of today, which is much less as compared to the national average of about 42 per cent. Not only the expansion of surface irrigation has been poor but the distribution of area under irrigation is also not the same across regions in the State (see, Tables 1.7 and 1.8).

However, as indicated earlier, the State has made rapid stride in transferring the irrigation system management to users (farmers) groups since mid-nineties. Since the performance of WUAs functioning at different locations using different sources of water is expected to be different because of heterogeneity in physical and other characteristics, we have selected altogether 10 WUAs having different physical and other characteristics. These 10 WUAs have been selected using the available Statistic as well as discussions with the officials of the irrigation department in Maharashtra State. Of the 10 WUAs, five have been selected from canal command areas, three from Lift Irrigation System (LIS) and two from Pani Panchayat scheme. Our aim is to see the functioning of WUAs at different locations of the canal command areas and therefore, one WUA has been selected from head reach and two each from middle and tail end of the canal command areas[12]. Similarly, three Associations established in lift irrigated areas using water from different sources namely river, well and back water have been selected. However, this same method has not been followed while selecting WUAs from Pani Panchayat since it mostly uses well water. The name, location and other details of these Associations selected for the study are presented in Table 2.1 in chapter two.

Table 1.9: Details of Selected Institutions and Sample Farmer Households

Particulars	*Canal System*				*Lift System*				*Pani Panchayat*	*Total*
	Head	*Middle*	*Tail*	*Total*	*River*	*Well*	*Backwater*	*Total*		
Number of Institutions Selected	01	02	02	05	01	01	01	03	02	10
Number of Sample Farmers Households										
Small	06	12	12	30	06	06	06	18	12	60
Medium	05	10	10	25	05	05	05	15	10	50
Large	04	08	08	20	04	04	04	12	08	40
Total	15	30	30	75	15	15	15	45	30	150
Percentage Distribution of Sample Farmers Households										
Small	10.0	20.0	20.0	50.0	10.0	10.0	10.0	30.0	20.0	100.0
Medium	10.0	20.0	20.0	50.0	10.0	10.0	10.0	30.0	20.0	100.0
Large	10.0	20.0	20.0	50.0	10.0	10.0	10.0	30.0	20.0	100.0
Total	10.0	20.0	20.0	50.0	10.0	10.0	10.0	30.0	20.0	100.0

Source: Designed by the authors.

The study has utilised data collected from both the sample farmers as well as from the records of all the 10 WUAs selected for the study. As regards the sample size, a total 150 sample farmers -75 farmers from canal irrigation system, 45 farmers from lift irrigation system and 30 from Pani Panchayat areas - have been selected randomly for the Study. Out of the 75 farmers selected from canal irrigation system, 15 farmers have been selected from WUA functioning in the head reach of the canal command, 30 from the middle region and another 30 from the tail end of the canal command area. Similarly, from lift irrigation system, 15 farmers have been selected from each WUAs operating under river water, well water and backwater. A total 30 sample farmers have been selected from the two Pani Panchayat WUAs. While the sample farmers have been selected randomly using the list of farmers provided by each WUA, farmers have been selected from each size group namely small, medium and large as well. The details of sample selection from each WUA and from each farm size group have been presented in Table 1.9.

The major objective of the study is to find out the overall impact of WUAs selected from different locations on various parameters. Therefore, field level data has been collected not only from the sample farmers but also from each selected WUAs. Data on water price, financial recovery, cropping pattern and other related information have been collected from the office records of each WUA to study the impact of it on various parameters. In order to find out the ground level impact of WUAs on different parameters, information on different items has been collected from the sample farmers.

Organisation of the Study

The study is organised into eight chapters. Second chapter of the study presents the profile of ten Water Users' Associations selected from different locations. The socio-economic characteristics of the 150 sample households selected for the study are presented in chapter three. The role of various functionaries and socio economic groups in the functioning of the Water Users' Association is presented in chapter four. To what extent the devolution of power to the Water Users' Association has taken place in the State is also analysed in the same chapter. The impact of Water Users' Association on cropping pattern, irrigated area, input use and productivity of crops is analysed in chapter five. The contribution of WUAs to the economic development of the selected areas as well as the impact of water users' association on efficiency, equity and social cohesion is discussed in chapter six. Problems faced by the sample farmers selected from various Water Users' Associations as well as their assessment on the success of the users' associations are presented in chapter seven. The last chapter presents the findings and recommendations of the study.

Notes

1. In addition to this, about Rs. 89,816 crore have been allocated for the irrigation development during the Tenth Plan Period (2002-07). If we include this, the total investment in irrigation sector would go up to Rs. 2454.42 billion.
2. Detailed analysis about the deteriorating financial performance of India's irrigation sector can be seen from GOI, (1992); Gulati, et al., (1994); Saleth, (1996); Deshpande and Narayanamoorthy, (2001); CWC, (2004).
3. The Command Area Development Programme (CADP) was introduced during the fifth plan period with an objective to increase over all efficiency of water use in the canal command areas by bridging the gap between potential created and utilised. As of 2001-02, about Rs. 6838 crore has been spent under this account. However, the gap between potential created and utilised has not declined after the introduction of CADP, but rather widened substantially.
4. How to form users' group and what kind of powers should be given to farmers in order to increase the participation of farmers in irrigation management have been clearly explained in Vaidyanathan committee report on pricing of irrigation water (GOI, 1992).
5. Unlike the formal WUAs established in canal command areas, the Pani Panchayat Institution is formed with certain novel principles which were not imposed by any external agents but emerged out of the discussions with the farmers keeping in view the water availability and potential for rainfed cultivation (Deshpande and Reddy, 1990; Thakur and Pattnaik, 2002; Keremane, et al., 2006). The principles followed in Pani Panchayat are given below:

(a) Only group schemes of minor irrigation would be taken up. This involves a group (community) and the community involvement is essential for any developmental process.

(b) Water would be supplied based on number of members in the family and not based on land owned by the members. Each member of the family would be entitled to get half an acre of irrigation with an upper ceiling of 2.50 acres for a household. This principle was followed to bring equity in water sharing.

(c) The members would have the rights to irrigation, but the water rights are not attached to the land. If a member sells his/her land the water rights are not transferred to the new owner. This would arrest the transfer of lands from the poor farmers to the richer ones due to the economic compulsion.

(d) The total cost of the scheme would be spread between the beneficiaries, GGP and subsidy from State government or bank loan in the ratio of 2:4:4. The beneficiaries would be fully responsible to administer and operate the scheme. This would ensure the involvement of the community both because of their personal investment and direct control on the scheme.

(e) Water-intensive crops such as sugarcane, paddy, and banana should not be cultivated by the members. This would enhance the spread of water and increase labour intensity in the region.

(f) The water rights (half an acre per capita) are also given to the landless agricultural labourers, who are working with the members in the scheme. These water rights would be transferred along with the labourer, where he/she chooses to work.

6. In the north-western part of Maharashtra, a communal irrigation management system prevails known as *phad* system. This system came to existence in the early *16th* century. *Phad* irrigation system is the oldest irrigation system practiced in Nasik and Dhule districts in the valley of Mosam and Panzara rivers, for over two centuries. The *phad* system is a most economical method of management of irrigation water. The system is based on the co-operation of farmers whose fields are to be irrigated. For details about working nature of *phad* system see, http://www.sristi.org.
7. Malgujari Tanks in Vidarbha region of Maharashtra were constructed nearly about two centuries back. These tanks are operated and managed by the farmers themselves. Though we do not have any information about the present performance of Malgujari tanks, at present, 325 malgujari tanks are in operation which covers 6771 hectares of area under the project Bagh, Itiyadoh, Ghodazari, Naleshwar, Asounlamendha, Chargaon and Chandai. For details see, http://waterconserve.maharashtra.gov.in
8. More details about the formation of WUAs and related issues can be seen from www. mahagovid.org
9. The complete Act on MMISF enacted by the Government of Maharashtra is presented in Appendix I of this study.
10. The Government of Maharashtra has made policy decision to

handover the irrigation management to WUA in July 2001 itself. However, the State could bring an Act to legalise the same only in May 2005. The main objectives of transferring the system to WUAs are: "(a) to reduce the gap between irrigation potential created and actual area irrigated, (b) to increase efficiency of irrigation management for water user, (c) to restrict the expenditure on maintenance and repairs of irrigation system, (d) to take up repairs of irrigation system before handing over, (e) to use water more efficiently and (f) to recover water charges effectively" (GOM, 2006. p. 202).

11. The broad methodology of the study was originally framed by the coordinators of the study (Prof. Vasant P Gandhi and Dr. N. V. Namboodiri, Centre for Management in Agriculture, Indian Institute of Management, Ahmadabad), after having one-day workshop on the study proposal with the participating Agro-Economic Research Centers. Selection of WUAs and the size of sample farmers for the study are fixed differently for each participating AERCs, depending upon the local condition. Since Maharashtra State has WUAs in canal command areas, lift irrigated areas and Pani Panchayat areas, our study covers all three types of WUAs.
12. These five WUAs have been selected from different minors coming under the Kukadi Left Bank Canal, which is a part of Kukadi irrigation project. Kukadi system is one of the major irrigation projects of Maharashtra State, which has created irrigation potential of 1.345 lakh hectares as of June 2005. The total length of the canal system is 249 kilometers and the districts benefited from this project are Pune, Ahmednagar and Sholapur. For more details about Kukadi project see, www.secadapune.org.

2

PROFILE OF SELECTED WUAs

Though the present work is predominantly based on the data collected from the sample farmers selected from different Water Users' Associations located at different areas, we have also attempted to study the overall profile of the selected WUAs. The analysis on the profile of WUAs would reveal the past and present status of each Association in terms its membership, area irrigated, cropping pattern, investment, pricing policy and other financial aspects, operation and maintenance of the irrigation channels, etc. Since each of the Association is selected from different locations that uses water from different sources, this analysis would also help us to understand the varied performance of each water users' association. As mentioned in the methodology section, altogether 10 water users' associations - five from Canal Irrigated Areas, three from Lift Irrigated Areas and two from Pani Panchayat Areas - have been selected for the present study. Let us now study each Association in detail.

Water Users Association – Canal (Head)

The performance of WUAs working in different locations of the canal command areas is expected to be different because of differences in quality of water supply and other reasons. Therefore, it is better to study each Association separately so as to understand how each one is performing. Out of five Water Users' Associations selected from the canal irrigated areas, two each have been selected from middle and tail end of the canals and one from head reach of the canal command areas.

Ganesh Water User Association

The Association which is selected from the head reach of canal command area is called as Shri Ganesh Water Users Association. This is

Table 2.1: Location of WUAs Selected for the Study

Irrigation System		*Name of Institution*	*Village / Taluka / District*
Canal System	Canal – Head (WUA_{CH})	1. Shri Ganesh Water Users Association, Mangrul (Minor No. 8 of Kukadi Left Bank Canal - KLBC)	Mangrul Ta. - Junner Dist. - Pune
	Canal – Middle (WUA_{CM})	2. Biroba Water Users Association (Cooperative), Ralegaon Therpal (Minor No. 24 of KLBC)	Ralegaon Thepal Ta. – Parner Dist. - Ahmednagar
		3. Shivshakti Co-operative Water Users Association, Nighoj (Minor No. 20 of KLBC)	Nighoj Ta. – Parner Dist. - Ahmednagar
	Canal – Tail (WUA_{CT})	4. Gangotri Co-operative Water Users Association, Belwandi (Minor No. 40 of KLBC)	Belwandi Ta. – Shrigonda Dist. - Ahmednagar
		5. Khandobaraya Water Users Association, Yelapane (SM 1 & 2, Minor No. 38 of KLBC)	Yelapane Ta. – Shrigonda Dist. - Ahmednagar

Lift System	Lift system – River (WUA_{LR})	6. Shri Chintamani Co-operative Water Users Association, Tarharwadi (Ghod river)	Tarharwadi (Avsari bk) Ta. – Ambegaon Dist. - Pune
	Lift system – Well (WUA_{LW})	7. Shri Shambhu Co-operative Water Users Association, Pimpalgaon (Khadaki) (Bank of Ghod river)	Pimpalgaon (khadaki) Ta. – Ambegaon Dist. - Pune
	Lift system - Backwater (WUA_{LBW})	8. Shri Wardhani Water Users Association, Takrarwadi (Ujani Dam- Bhima river)	Takrarwadi Ta. – Indapur Dist. - Pune
Pani Panchayat	Pani Panchayat (WUA_{PP})	9. Renuka Lift Irrigation Scheme, Mahoor (Minor Tank – Vir Dam)	Mahoor Ta. – Purndar Dist. - Pune
		10. Babdeo Irrigation Scheme, Shindewadi (Minor Tank – Nira Basin West)	Shindewadi Ta. – Purndar Dist. - Pune

Source: Field survey data.

Table 2.2: Number of Members in Each WUA by Land Size

Sr. No.	*WUAs*	*Total membership in the Institution*			*Membership by land Ownership*				*Formal terms and conditions to become a member of the Institution*
		Male	*Female*	*Total*	*Large and Medium Farmers (> 10 acres)*	*Small Farmers (2.5 to 10 acres)*	*Marginal Farmers (<2.5 acres)*	*Landless*	
1.	WUA_{CH}	125	5	130 (80)	0	88	42	0	• Own land in village • Share deposit Rs. 50/-
2.	WUA_{CM}	102	10	112 (78)	8	76	28	0	• Own land in village • Share deposit Rs. 61/-
3.	WUA_{CM}	186	21	207 (152)	17	80	110	0	• Own land in village • Share deposit Rs. 100/-

4.	WUA_{CT}	104	10	114 (114)	25	60	29	0	• Own land in village • Share deposit Rs. 51/-.
5.	WUA_{CT}	112	10	122 (122)	20	72	30	0	• Own land in village • Share deposit Rs. 51/-
6.	WUA_{LR}	138	2	140 (64)	6	51	83	0	• Own land in village • Required deposit should be paid.
7.	WUA_{LW}	75	5	80 (60)	5	34	41	0	• Own land in village • Required deposit should be paid

8.	$WUAL_{BW}$	25	4	29 (26)	5	10	14	0	• Own land in village • Required deposit should be paid
9.	WUA_{PP}	46	2	48 (53)	4	30	12	0	• Should be agreed to Pani Panchayat rules
10.	WUA_{PP}	26	3	29 (52)	0	0	29	0	• Should be agreed to Pani Panchayat rules

Note: Figures in brackets are number of members at the time of formation of WUA.
Source: Field survey data

Table 2.3: Basic Details of WUAs Selected for the Study

Sr. No	*WUAs and Year of establishment*	*No of Farming Households*	*Nature of activities*	*Signed MOU with Govt.*	*Nature of Power given to Manage Irrigation system*	*Any subsidy / funding from Govt.*	*Water rate (Rs / ha)*
1.	WUA_{CH} 1996	13	• Take water on volumetric basis and distribute to farmers • Repair works • Fixing and collection of water rates	Yes	• Regular water distribution. • Recovery of water charges. • Repair works.	Yes Rs. 84000 from State Government for maintenance and repair	As per government rate
2.	WUA_{CM} 1994-95	112	• Take water on volumetric basis and distribute to the farmers. • Water charges as per government rates • Repair works	Yes	• Regular water distribution. • Recovery of water charges. • Repair works.	Yes 1st year Rs.100/ha, 2nd year Rs.100/ha, 3rd year 75/ha, after that 20% discount on bill and 5% discount if water charges paid in time	As per government rate

3.	WUA_{CM} 1994	207	• Take water on volumetric basis and distribute to the farmers. • Water charges as per government rates • Repair works	Yes	• Regular water distribution • Recovery of water charges. • Repair works.	Rs. 100/ha (on irrigated area)	20% more than Government rate
4.	WUA_{CT} 1996	114	• Take water on volumetric basis and distribute to the farmers. • Water charges as per government rates • Repair works	Yes	• Regular water distribution. • Recovery of water charges. • Repair works.	Rs. 32000/- last year	As per government rate
5.	WUA_{CT} 1999-00	122	• Take water on volumetric basis and distribute to the farmers. • Water charges as per government rates • Repair works	Yes	• Regular water distribution. • Recovery of water charges. • Repair works.	Rs. 54000/- from Irrigation Department	As per government rate

6.	WUA_{LR} 1975	140	• Proper maintenance/ repair. • Collection of water charges. • Monitoring use of water. • Planning for release of water, etc.	Yes	• The Society charges water rates on crop area basis. • The society has the power to charge service fee to cover part of the management cost.	No	Sugarcane - 650/-Other crops - 450/-
7.	WUA_{LW} 1986	80	• Proper maintenance/ repair • Collection of water charges • Monitoring use of water. • Planning for release of water, etc.	Yes	• The Society charges water rates on crop area basis. • The society has the power to charge service fee to cover part of the management cost.	No	Sugarcane- 600/-, Other crops - 400/-

8.	$WUAL_{BW}$ 1981	29	• Proper maintenance/ repair. • Collection of water charges. • Monitoring use of water. • Planning for release of water, etc.	Yes	• The Society charges water rates on crop area basis. • The society has the power to charge service fee to cover part of the management cost.	Yes Rs. 48000/- at the time of formation	Sugarcane - 1000/- Other crops - 200/-
9.	WUA_{PP} 1982	48	• Water distribution. • Maintenance of electric motor and pipeline. • Collect shares from members	–	• As per Pani Panchayat objectives	50% subsidy at the time of formation	Rs. 200-300/-
10.	WUA_{PP} 1986	29	• Water distribution. • Maintenance of electric motor and pipeline. • Collect shares from members	–	• As per Pani Panchayat objectives	In 1980 50% subsidy on expenditure	Rs. 200-300/-

Source: Field survey data.

located in Mangrul village of Junner taluka, Pune district. This society was established in 1996 with the initiation of some progressive farmers in the village and also with the support of the officials belonging to the state irrigation department. This WUA was started with 80 member farmers, but now the same has increased to 130. This increase in membership happened partly because of the performance of WUA and also due to the condition imposed by the state government.[1] Out of 130 members, 125 are male members and remaining five are from female group. Surprisingly, large-size farmers are completely absent in the association, whereas marginal and medium-size farmers account for 32 per cent and 68 per cent respectively in the Association (see, Tables 2.1, 2.2 and 2.3). Farmers must have own land in the village/command area and must pay a nominal membership fee to become a member of WUA.

Water distribution as well as decision regarding the cropping pattern for each season is decided by the members themselves. According to the information received from the office of WUA, the formation of WUA has made considerable impact on the area under cultivation as well as on irrigated area. While the gross cultivated area has increased by 326 percent between 1996 and now, the area under irrigation has increased by 200 percent during the same period. Proper monitoring and supervision of water supply in the area covered by WUA reported to have helped to increase the irrigated area along with gross cropped area (see, Tables 2.4 and 2.5).

Though WUA has the right to decide the water rates for each crop and each season, the Association has been following the water rates decided by the State government. Water rates have been collected from the farmers at the end of each season, which is ultimately paid to the irrigation department as per the regulation of the government. Since water supply would be stopped for those farmers who do not pay the water charges in time, defaulters are generally not seen in this Water Users' Association. The overall financial status of WUA is reported to be satisfactory (see, Tables 2.6, 2.7 and 2.8).

The distribution of water and other related works have been regularly monitored by WUA, for which two employees have been appointed. This WUA has not faced any serious problem in executing the task which is assigned to it so far. However, the office bearers of the users association suggested that timely availability of water and uninterrupted support from the State government for carrying out the operation and maintenance work are essential to sustain the performance of WUA.

Table 2.4: Functions and Activities of the Selected WUAs

Sr No	*WUAs*	*Capital Investment (Rs)*	*Mainte-nance / Repair (Rs.)*	*Water allocation / distribution*	*Pricing*	*Cropping pattern decided by*	*Fee Collection*	*Monitoring and Control*	*Relationship with Government*
1.	WUA_{CH}	7000	10000	Tail to head distribution	Government rates	Farmers	Collected every season	Appointed one person by society	Smooth
2.	WUA_{CM}	6000	11500	Tail to head distribution	10 % more than Government rate	Farmers	Collected every season	Appointed one person by society	Smooth
3.	WUA_{CM}	5000	2000	Tail to head distribution	Government rate	Farmers	Collected every season	Appointed one person by society	Smooth
4.	WUA_{CT}	11200	16250	Head to tail distribution	Government rate	Farmers	Collected every season	Appointed one person by society	Smooth
5.	WUA_{CT}	5000	10000	once tail and once head rotation	Government rate	Farmers	Collected every season	Appointed one person by society	Smooth

6.	WUA_{LR}	9500	6000	Traditional System per farm (Rotation)	Cropwise/ ha (Rs. 80/-per acre/ irrigation)	Farmers	Collected every month as per rotation	Appointed one person (farmer) by society	Good
7.	WUA_{LW}	400000	7900	No measuring device, Rotation System	Cropwise/ ha (Rs. 80/-per acre/ irrigation)	Farmers	Collected every month as per rotation	Appointed one person (farmer) by society	Good
8.	WUA_{LBW}	125000	19000	No Measuring device, Rotation System	Cropwise/ ha (Rs. 80/-per acre/ irrigation)	Farmers	Collected every month as per rotation	Appointed one person (farmer) by society	Good
9.	WUA_{PP}	150000	11250	from tail to head distribution	On acre basis (Rs. 80/-per acre/ irrigation)	Farmers	Every month/ irrigation	Appointed one person (farmer) by society	Good
10.	WUA_{PP}	170000	7125	from tail to head distribution	Cropwise/ acre basis	Farmers	Every month/ irrigation (at the beginning of season)	Hired person from outside	Smooth

Source: Field survey data.

Table 2.5: Command Area, Gross Cropped Area and Irrigated Area of the Selected WUAs

Sr No.	*WUAs No*	*Command Area (ha)*	*Total Cultivated Area in the Command (ha)*					*Total Irrigated Area (ha)*				
			Total	*Kharif*	*Rabi*	*Summer*	*Perennial*	*Total*	*Kharif*	*Rabi*	*Summer*	*Perennial*
At the Time of the Formation of WUAs												
1.	WUA_{CH}	65.0	75.0	20.0	40.0	15.0	0.0	75.0	20.0	40.0	15.0	0.0
2.	WUA_{CM}	315.0	357.0	145.0	152.0	60.0	0.0	236.9	86.1	91.1	59.7	0.0
3.	WUA_{CM}	321.0	115.0	22.0	61.0	17.0	15.0	115.0	22.0	61.0	17.0	15.0
4.	WUA_{CT}	305.0	421.0	175.0	190.0	56.0	0.0	256.0	55.0	145.0	56.0	0.0
5.	WUA_{CT}	915.0	585.0	150.0	325.0	90.0	20.0	435.0	90.0	235.0	90.0	20.0
6.	WUA_{LR}	100.8	158.0	97.6	60.0	0.0	0.4	158.0	97.6	60.0	0.0	0.4
7.	WUA_{LW}	42.4	62.1	40.0	21.1	0.0	1.0	62.1	40.0	21.1	0.0	1.0
8.	WUA_{LBW}	48.0	67.4	26.3	15.5	4.7	20.9	67.4	26.3	15.5	4.7	20.9
9.	WUA_{PP}	150.0	27.0	15.0	12.0	0.0	0.0	2.0	2.0	0.0	0.0	0.0
10.	WUA_{PP}	32.0	64.0	26.0	26.0	12.0	0.0	64.0	26.0	26.0	12.0	0.0
At Present												
1.	WUA_{CH}	220.0	320.0	110.0	120.0	90.0	0.0	190.0	60.0	90.0	40.0	0.0
2.	WUA_{CM}	315.0	402.0	160.0	175.0	67.0	0.0	252.0	65.0	120.0	67.0	0.0
3.	WUA_{CM}	321.0	310.0	80.0	150.0	55.0	25.0	310.0	80.0	150.0	55.0	25.0
4.	WUA_{CT}	305.0	465.0	185.0	210.0	70.0	0.0	380.0	115.0	185.0	80.0	0.0
5.	WUA_{CT}	915.0	653.0	225.0	355.0	40.0	33.0	530.0	200.0	290.0	7.0	33.0

6.	WUA_{LR}	103.7	218.3	90.5	70.0	45.0	12.8	218.3	90.5	70.0	45.0	12.8
7.	WUA_{LW}	52.5	110.5	41.5	30.0	28.0	11.0	110.5	41.5	30.0	28.0	11.0
8.	WUA_{LBW}	55.0	81.5	29.0	20.1	7.6	24.8	81.5	29.0	20.1	7.6	24.8
9.	WUA_{PP}	150.0	135.0	70.0	50.0	15.0	0.0	34.0	6.0	24.0	4.0	0.0
10.	WUA_{PP}	12.0	30.0	12.0	12.0	6.0	0.0	30.0	12.0	12.0	6.0	0.0
% Change												
1.	WUA_{CH}	233.5	326.7	450.0	200.0	500.0	_	153.3	200.0	125.0	166.7	_
2.	WUA_{CM}	0.0	12.6	10.3	15.1	11.7	_	6.4	-24.5	31.7	12.2	_
3.	WUA_{CM}	0.0	169.6	263.6	145.9	223.5	66.7	169.6	263.6	145.9	223.5	66.7
4.	WUA_{CT}	0.0	10.5	5.7	10.5	25.0	_	48.4	109.1	27.6	42.9	_
5.	WUA_{CT}	0.0	11.6	50.0	9.2	-55.6	65.0	21.8	122.2	23.4	-92.2	65.0
6.	WUA_{LR}	2.8	38.1	-7.3	16.7	_	3100.0	38.1	-7.3	16.7	_	3100.0
7.	WUA_{LW}	23.7	77.9	3.6	42.2	_	1000.0	77.9	3.6	42.2	_	1000.0
8.	WUA_{LBW}	14.6	21.0	10.4	29.8	61.7	18.7	21.0	10.4	29.8	61.7	18.7
9.	WUA_{PP}	0.0	400.0	366.7	316.7	_	_	1600.0	200.0	-	_	_
10.	WUA_{PP}	-62.5	-53.1	-53.8	-53.8	-50.0	_	-53.1	-53.8	-53.8	-50.0	_

Source: Field Sourcies Data

Table 2.6: Financial Aspects of the Selected WUAs

WUAs	WUA_{CH}		WUA_{CM}		WUA_{CM}	
	ATF	Now	ATF	Now	ATF	Now
Water Rate payable (Rs/ ha)	As per Govt. rates	As per Govt. rates	As per Govt. rates	Govt. rates + 10% extra	As per Govt. rates	As per Govt. rates
Water Rate collected per ha (or the unit)	As per Govt. rates	As per Govt. rates	As per Govt. rates	As per Govt. rates	As per Govt. rates	As per Govt. rates
Total amount to be paid by the farmers	11894	14000	55000	85000	80000	85000
Total amount collected from farmers (Rs.)	2900	14000	43200	85000	80000	85000
Arrears (Rs)	8904	-	-	-	-	-
Payment Balance to Government	Nil	Nil	8000	Nil	Nil	15000
Capital Investment (Rs)	7000	1480	6000	11000	5000	7000
Any other Revenues (Rs)	-	-	-	-	-	-
Expenditure (Rs)	6421	14000	29000	34916	5000	8500
Profit/Loss (Rs)	+5634	+6400	+11456	+11615	+9000	+7500

WUAs	WUA_{CT}		WUA_{CT}		WUA_{LR}		WUA_{LW}	
	ATF	*Now*	*ATF*	*Now*	*ATF*	*Now*	*ATF*	*Now*
Water Rate payable (Rs/ ha)	As per Govt. rates	As per Govt. rates	As per Govt. rates	As per Govt. rates	40	450-650	35	400-600
Water Rate collected per ha (or the unit)	As per Govt. rates	As per Govt. rates	As per Govt. rates	As per Govt. rates	40	450-650	35	400-600
Total amount to be paid by the farmers	56808	90000	18000	99657	3600	2000	7900	2700
Total amount collected from farmers (Rs.)	37050	90000	16500	99657	850	2000	1250	2700
Arrears (Rs)	-	-	-	-	-	-	-	-
Payment Balance to Government	32688	46000	Nil	Nil	2750	Nil	6650	Nil
Capital Investment (Rs)	11200	15000	5000	10000	95000	-	400000	-
Any other Revenues (Rs)	-	-	-	-	-	-	-	-
Expenditure (Rs)	91648	53771	56108	65000	2900	22316	4250	16800
Profit/Loss (Rs)	-91848	-12565	-40933	-12000	-	-	-	-

WUAs	WUA_{LBW}		WUA_{PP}		WUA_{PP}	
	ATF	*Now*	*ATF*	*Now*	*ATF*	*Now*
Water Rate payable (Rs/ha)	100-400	138-1000	Govt. rate	200-300	Govt. rate	200-300
Water Rate collected per ha (or the unit)	100-400	138-1000	Govt. rate	200-300	Govt. rate	200-300
Total amount to be paid by the farmers	4250	2300	1040	2100	3350	2370
Total amount collected from farmers (Rs.)	450	2300	590	2100	650	2370
Arrears (Rs)	-	-	-	-	-	-
Payment Balance to Government	3800	Nil	2300	Nil	2700	Nil
Capital Investment (Rs)	125000	-	150000	-	170000	-
Any other Revenues (Rs)	-	-	-	-	-	-
Expenditure (Rs)	3290	18600	1200	20000	1900	18200
Profit/Loss (Rs)	-	-	-	-	-	-

Note: ATF – at the time of formation of WUA.
Source: Field survey data.

Table 2.7: Assessment about the Success and Financial Health of WUAs

Sr. No.	*WUAs*	*Assessment about the success of WUAs*	*Assessment about the financial health of WUAs*
1.	WUA_{CH}	Successful	Satisfactory
2.	WUA_{CM}	Satisfactory	Satisfactory
3.	WUA_{CM}	Satisfactory	Satisfactory
4.	WUA_{CT}	Satisfactory	Satisfactory
5.	WUA_{CT}	Successful	Satisfactory
6.	WUA_{LR}	Successful	Strong
7.	WUA_{LW}	Successful	Strong
8.	WUA_{LBW}	Successful	Satisfactory
9.	WUA_{PP}	Successful	Satisfactory
10.	WUA_{PP}	Poor	Poor

Source: Field survey data.

Water User Association – Canal (middle)

Two Water Users' Associations have been selected from the middle reach of the canal command area for the study, which are called as Biroba Water Users' Association and Shivshakti Cooperative Water Users' Association.

Biroba WUA:

Biroba WUA is located in the village of Ralegaon Thepal, Parner taluka of Ahmednagar district. This was started during 1994-95 with the initiation of the irrigation department officials. The total membership of the Association was 78 at the time of starting, which now increased to 112. Female members account for around nine per cent of the total membership. Of the total farmers, just over seven per cent of the farmers are from large-size category, having land over four hectares and the remaining farmers are marginal and small size group, having less than four hectares of land. Farmers having own land in the command area are allowed to become a member of the Association.

Moderate increase in cropped area as well as in irrigated area is noted after the establishment of WUA. While the cropped area increased about 12 per cent between 1994-95 and now, the irrigated area has

Table 2.8: Major Problems and Suggestions on Various Aspects of WUAs

Sr No.	*WUAs*	*Major Problems faced since inception of WUA*	*Is devolution of power to WUA adequate?*	*Major factors / requirements for the Success of WUA*
1.	WUA_{CH}	No Problem	Yes	Regular recovery is essential Timely water availability is essential.
2.	WUA_{CM}	Poor maintenance of main and sub-channels.	Yes	Recovery on time is essential. Timely water availability is essential. Advance information about water availability to plan crop pattern
3.	WUA_{CM}	Because of non availability of information on water supply rotation, not able to plan cropping pattern in advance.	Yes	Recovery on time is essential. Timely water availability is essential.
4.	WUA_{CT}	During summer, water shortage occurs.	Yes	Government support is needed Salaries of Secretary and Patwari should be given by Government.
5.	WUA_{CT}	Due to poor maintenance of distributory/ sub-channels, huge loss of water Sometime tail end farmers don't get water Repair works of channels are not possible by WUA.	Yes	Government subsidy is needed Repair work of channels should be carried out by government. Capital assistance should be provided by government.

6.	WUA_{LB}	Lack of funding from the government. Erratic electric supply	Yes	Government support needed for maintenance work Subsidy on electricity supply
7.	WUA_{LW}	Lack of funding from the government. Erratic electric supply	Yes	Government support needed for maintenance work Subsidy on electricity supply
8.	WUA_{LBW}	Occasionally problem with water supply (during repair works) Erratic electric supply	Yes	Government support needed for maintenance work Subsidy on electricity supply
9.	WUA_{PP}	Problems arise when electricity bill not paid. Shortage of water in summer Erratic electric supply	Yes	Government support needed for maintenance work Subsidy on electricity supply
10.	WUA_{PP}	Due to unpaid-electricity bill, scheme may be closed down soon. Erratic electric supply Shortage of water in summer.	Yes	Government support for maintenance work Subsidy on electricity supply Due to voltage problem, electric motor pump gets in to problem resulting in high cost of repair.

Source: Field survey data.

increased by about six per cent during the same period. Though area under irrigation has not increased appreciably over the period, the Association is able to reduce the problems related to water distribution and supply through consensus among the members, which was not possible prior to the formation of WUA in this area.

The water rates have been fixed by following the rates announced periodically by the government. Since the Association is able to collect the water rates from farmers in time, the financial status of the Association is reported to be satisfactory. The members of management committee feel that the devolution of power given to them is adequate. However, the office bearers of the users' Association suggested that the advance information about water availability and timings of release of water would help to improve the performance of WUA to a greater extent.

Shivashakti Corporative WUA

The other WUA selected from the middle reach of the canal command area is called Shivshakti Cooperative Water Users' Association, which is located in Nighoj Village, Parner Taluka, Ahmednagar district. The establishment year of this Users Association was 1994. Frequent water scarcity problem encountered by the farmers of this village appear to be the main reason behind the formation of the Association. Similar to other WUA selected from middle reach of canal command area, the number of membership increased from 152 at the time of establishment to 207 at present. This society has both male and female members. Of the total member farmers, large-size farmers (above 4 ha) account for about eight per cent and the remaining are marginal and small size farmers.

As per the information supplied by the office of the Association, both the cropped area and irrigated area have increased considerably after the formation of Users' Association (see, Tables 2.1 to 2.8). This increase seems to have been achieved because of proper monitoring and supervision of the members who manage the water users' Association.

Unlike other WUAs discussed above, Shivshakti Cooperative WUA charges water rates 20 per cent above the rate fixed by the government for various crops. This has been done specifically to strengthen the financial status of the society as well as to improve the operation and maintenance of the channels which are being used for taking water upto the farmers' field. Presently this society runs with some profit. The management committee of Users' Association also feels that the devolution of power given to it is adequate to carry out the specified works and objectives. However, the office bearers feel that because of non-availability of information on the timings of water supply from the irrigation department, the users are not able to decide the cropping pattern in advance.

Water User Association – Canal (Tail end)

Two Water Users' Associations working at the tail end of the canal irrigated areas have also been selected for the study. There names are Gangotri Cooperative Water Users Association and Khandobaraya Water Users' Association. Gangotri Association is located in Belwandi village, Shrigonda taluka of Ahmednagar district, whereas the Khandobaraya Association is established in Yelapane village in Ahmednagar district. Both are established relatively recently, because of the persuasion by the officials from irrigation department. Unlike the changes noticed in WUAs located at the head reach of the canal command area, the expansion of gross cultivated area and irrigated area is not substantial since the establishment of these two WUAs. Though many earlier studies have reported that the tail-end farmers of the canal command area often encounter with water scarcity problem[2], we did not receive any serious complaint about the water scarcity from the office bearers of the Users' Association. In fact, the water scarcity problem appears to have reduced substantially after the establishment. The financial status of the two WUAs reported to be satisfactory. Their overall status is presented in Tables 2.1 to 2.8.

Lift Irrigation WUAs

The Water Users' Associations that are functioning in the lift irrigated areas are in many ways different from the one that is working in the canal irrigated areas. WUA can be formed without investing huge capital in the canal irrigated areas, whereas huge capital investment is necessary to form WUAs in the lift irrigated areas. The water pricing policy, cropping pattern and the operation and management aspects of lift irrigation are also totally different from the Water Users' Association established in the canal irrigated areas. Therefore, in order to understand the overall working nature of WUAs established in the lift irrigated areas, three different WUAs using water from river, well and backwater have been selected for the study along with other. Let us look at the overall working condition of the WUA functioning in the lift irrigated areas.

Unlike the Canal Water Users' Association, all the three water users associations working in the lift irrigated areas are relatively old. Of the three WUAs, the Association working under river system was established during 1975, well system was in 1986 and the system that works in back water was established in 1981. The number of members registered with all three WUAs have also increased over the years.

Pricing System

The water pricing system followed by the WUA established in lift irrigated area is different from the canal irrigated area. The water price fixed for water-intensive crops like sugarcane is relatively higher than that of other crops. It appears that each WUA decides the price for water for different crops on its own depending upon the operation and maintenance costs, demand and supply of water, cropping pattern, etc. Since lift irrigation requires relatively large capital investment, all the three WUAs have invested huge amount of money to establish their society, which is also evident from the data supplied by them (see, Table 2.1 to 2.8). However, as per the data provided by the society office, the financial status of the lift irrigation society established in the area where backwater of the dam is used as a source of water is not satisfactory. Substantial increase in operation and maintenance expenditures (which includes electricity charges) is cited as the main reason for the poor financial status of the lift irrigation society.

Pani Panchayat (Water Council)

Two Water Users' Associations working in Pani Panchayat areas have also been selected for the study. The working nature of the Associations established in Pani Panchayat areas are totally different from the these discussed above. As mentioned earlier, these Users' Associations are established in the rainfed areas with the aim to protect the livelihood of the people often affected by droughts. This institution is not promoted by the State agencies, but established by an NGO with the voluntary support from the users' group.

During 1972-73, Maharashtra had to face one of the worst droughts in its history. Purandhar taluka of Pune district, one of the drought prone talukas even today, had to suffer both for water and food during that drought. At this juncture, a young engineer by name Shri Vilasrao Salunke, working on some committee of the Government noted that the solutions provided through government are not going to solve the problems of drought faced by the rural population permanently. He thought that the most important contributing resource of rural India is water, which should be utilised optimally to remove the suffering of people from droughts. In order to utilise the natural resources, a Trust

(NGO) with a name Gram Gourav Pratishthan (GGP) was started with the donation from individuals and private agencies. Initially, GGP took a 16 hectares of degraded temple lands for lease, which were on the hill slopes of Naygaon village in Purandhar taluka. The land serves as a natural watershed. There is a natural basin in the down reaches of a capacity of one million cubic feet, which was repaired and made as a percolation tank. An open well was dug at the down stream base of percolation tank to irrigate the lands leased by GGP. The water and soil conservation experiments provided irrigation for eight months. These activities resulted in 9.60 ha of irrigated area and 2.40 ha under tree crops. The cropping pattern changed from low value crops to relatively high value crops in this area and the yield rate was also improved substantially because of the impact of Pani Panchayat (see, Deshpande and Reddy, 1990).

Having seen the effect of this experiment, farmers belonging to Naygaon and peripheral areas realised that there is great hope for rainfed agriculture with the marginal support of irrigation. The experiment brought the farmers together voluntarily and the discussion resulted in the formation of Pani Panchayat. The principals followed under Pani Panchayat were not imposed by GGP or by any external agent, but emerged out of the discussions with the farmers keeping in view the water availability and potential for rainfed cultivation (Deshpande and Reddy, 1999; Thakur and Pattnaik, 2002; Keremane, et al., 2006).

It is evident from Tables 2.1 to 2.8 that the present status of Pani Panchayat system is not very encouraging. Because of inadequate supply of water, restrictions for cultivating high value crops, etc., many farmers have opted out from Pani Panchayat institutions. The cropped area irrigated has increased substantially immediately after the establishment of Pani Panchayat, but the irrigated area has come down in recent years. Of the two WUAs selected from Pani Panchayat areas, Renuka Lift Irrigation Scheme located in Mahoor Village of Purandar taluka, Pune district, still seems to be working without any financial losses.

To sum up, the profile of the ten WUAs selected for the analysis shows diverse results. The overall working condition of WUAs established in the canal irrigated areas and lift irrigated areas seem to be better. But the same is not fully true with WUAs established in Pani Panchayat areas. Considerable increase in gross cropped area and irrigated area has been reported due to the formation of Water Users' Association. The financial status of the two WUAs – one from lift WUA and another one from Pani

Panchayat – seems to be in bad condition because of increased operation and maintenance expenditures.

Notes

1. As per the Maharashtra Management of Irrigation System by Farmers Act, 2005, water will be supplied only to water users associations and not to individual farmers.
2. The problems encountered by the tail enders of the canal command area in States like Tamil Nadu, Karnataka and Maharashtra are clearly documented by Rajagopal, et al., (2002).

3

FEATURES OF SURVEY HOUSEHOLDS

The socio-economic characteristics of the households generally play a major role in decision making process in various aspects of the rural economy. Therefore, in this Chapter, we have looked at some of the characteristics of sample households. Specifically, we have looked at the level of education of the respondents, farming experience, family size, position in the village, position in their respective Water User Associations (WUAs) including their participation level, land holding status as well as the irrigation availability of the sample households in the command area where the survey was carried out. Since the sample farmers have been selected from 10 different WUAs, working under canal irrigation system, lift irrigation system and Pani Panchayat areas, the characteristics of the households are expected to be different across different locations. Therefore, we have discussed the characteristics of the households separately taking into account each WUA.

Overall Characteristics of Households from WUAs

As mentioned earlier, we have selected altogether 10 WUAs working under canal irrigation system, lift irrigation system and Pani Panchayat areas. The total sample size of the study is 150, which consists of 56 small farmers, 50 medium farmers and 40 large size farmers. Though our objective is to study the characteristics of the households separately selected from different WUAs, let us first look at the overall characteristics of the households selected from all the 10 WUAs.

Table 3.1 presents the basic characteristics of all the 150 sample households selected for the study. Though one may theoretically expect better education and other characteristics among the large size farmers' households, our survey data does not show any vivid difference in the case of education and farming experience across different farm size groups. While the average score of education of the respondent varies from 2.50 to 2.90 among the three land size groups, the farming experience varies from about 24 years to 27 years. However, against our expectation, the family size of large farmers group is relatively larger (9.13) than that of other two groups of farmers' households namely small (6.58) and medium (7.52).

Table 3.1: Basic Characteristics of Sample Households – Results of All WUAs

Particulars	*Farm Size*			
	Small	*Medium*	*Large*	*Overall*
Number of respondents reported as head of household	56 (60)	44 (50)	38 (40)	138 (180)
Age of the head of household (years)	48.47	49.84	48.68	48.98
Education of the respondent (code)*	2.50	2.50	2.90	2.61
Farming experience (years)	25.28	27.54	24.20	25.75
Average size of family	6.58	7.52	9.13	7.57
Distance from nearest town/ city (km)	14.18	15.28	14.10	14.53
Position in the Village (Percentage to total sample)				
Head of households	93.33	88.00	95.00	92.00
Village leader	11.67	12.00	30.00	16.67
Position in Water User Association				
Non-member	0	0	0	0
Member	54	45	29	128
Managing committee member	4	3	4	11
Chairman	0	2	4	6
Vice-chairman	0	0	1	1
Secretary	1	0	0	1
Staff	1	0	0	1
Others	0	0	2	2

Participation in activities/ decision-making of WUA (Percentage)				
Very active	15.0	12.0	22.5	16.0
Active	60.0	66.0	60.0	62.0
Passive	25.0	22.0	17.5	22.0
None	0.0	0.0	0.0	0.0

Notes: Figures in brackets are total number of farmers selected from each size group. * - Education is given in code - 1 (No formal education); 2 (Primary); 3 (Secondary); 4 (Graduate) and 5 (above Graduate).
Source: Field survey data.

As regards the position of the respondents in the sample villages, our survey data shows that over 90 per cent of the respondents are the head of the households, which is almost same across all the three farm size categories. However, nearly one third of the respondents belonging to the large farm size groups are occupying the position of village leader, which is very less among the small and medium size group of farmers. Since most of the large size farmers are rich and high caste Hindu, they one way or the other could become village leader, which is a common phenomenon across different regions in India.

With regard to the position occupied by the respondents in WUAs, it is very evident from Table 3.1 that most of the respondents are just members of WUAs. However, it is interesting to note that nearly 23 per cent of the respondents belonging to the group of large farm size could occupy the honourable positions such as Chairman, Vice-Chairman and Managing Committee Members in different WUAs, which is very less among the small and medium size group of farmers. Since most of the farmers belonging to large size are rich, they are able to win the elections and occupy the respectable position in WUAs.

We have also looked at as to how enthusiastically different groups of farmers are participating in decision making process of WUAs. In order to know this, farmers were asked to indicate whether they participate (a) very actively, (b) actively, (c) passively and (d) none. Our survey reveals that an average only 16 percentage of the farmers expressed that they are very actively participating in the decision making process of WUAs. However, as expected, this percentage is positively correlated with farm size. That is, the level of active participation is found to be relatively higher among the large farm size group than that of the other two groups of farmers. Since the requirement of water for large farm size is very high, these farmers generally take more interest in the decision making process so as to get the water in time for crop cultivation. Farmers who expressed that they actively participate in the activities pertaining to WUAs account for over 62 per cent, which is almost same in all the three groups of farmers. Nearly one fourth of farmers belonging to small and medium group expressed that their participation in WUAs related activities are passive because of various reasons. On the whole, over 78 per cent of the farmers have expressed that they have either very actively or actively participated in various activities related to WUAs.

Characteristics of Households – Individual WUA

In the previous section, we have looked at the overall characteristics of the respondents belonging to all the three types of WUAs. The characteristics of the farmers' households are expected to be different with different WUAs. Therefore, in this section, we have looked at whether any distinct difference exists in the characteristics of the households selected from different WUAs namely those working under canal irrigation system, lift irrigation system and Pani Panchayat areas. Since the source of water used by each is different, we expect certain variations in the characteristics of households.

Tables 3.2, 3.3 and 3.4 separately present the characteristics of the sample households selected from WUAs working under canal system, lift irrigation system and Pani Panchayat. As seen earlier, there is so much difference in the average level of education of the respondents across all the three WUAs. The average education of the respondents varies from 2.50 to 2.70 among the three WUAs. In the case of age of respondents and farming experience, we could notice a vivid difference among the households belonging to different WUAs. For instance, the average age of the respondents selected from canal irrigated areas is relatively lower (43.16) as compared to the farmers selected from lift irrigated areas (55.98) and Pani Panchayat system (53.03). Similar to this, we could also notice some differences in the farming experience of the sample farmers. Though there are differences in the family size of the households selected different WUAs, family size is positively related with the farm size in all the three types of WUAs.

We could also see a distinct difference in the position of the respondents in their respective villages among the three groups of WUAs. Though over 90 per cent of the respondents are head of the households in all the three groups of WUAs, over 35 to 41 per cent of respondents belonging to large farm size are occupying the position of village leader in the canal as well as lift irrigation system. However, this is not true in

Table 3.2: Basic Characteristics of Sample Households - Canal Irrigation System

Particulars	*Farm Size*			
	Small	*Medium*	*Large*	*Overall*
Number of respondents reported as head of household	27 (30)	23 (25)	19 (20)	69 (75)
Age of the head of household (years)	43.03	44.60	41.55	43.16
Education of the respondent (code)	2.37	2.52	2.85	2.55
Farming experience (years)	23.03	23.52	19.45	22.24
Average size of family	7.03	6.80	9.20	7.53
Distance from nearest town/ city (km)	17.37	19.36	17.20	17.99
Position in the Village (Percentage to total sample)				
Head of households	90.00	92.00	95.00	92.00
Village leader	3.33	16.00	35.00	16.00
Position in Water User Association				
Non-member	0	0	0	0
Member	30	23	14	67
Managing committee member	0	1	3	4
Chairman	0	1	1	2
Vice-chairman	0	0	0	0
Secretary	0	0	0	0
Staff	0	0	0	0
Others	0	0	2	2
Participation in activities/ decision-making of WUA (Percentage)				
Very active	3.3	4.0	5.0	4.0
Active	63.3	72.0	80.0	70.7
Passive	33.3	24.0	15.0	25.3
None	0.0	0.0	0.0	0.0

Note and *Source*: Same as in Table 2.1.

those WUAs, which are working under Pani Panchayat areas.

Differences are also noticed in the position that the respondents occupy in their respective WUAs across the three groups of WUAs. Considerable percentage of farmers belonging to large farm size group are occupying some respectable position (either Chairman or Vice-Chairman or Managing Committee Member) in the WUAs working under canal and lift irrigation system. But, this is not true in the case of Pani Panchayat, where almost all the respondents are simple members of the Water Users' Association.

Whether or not any significant difference exists among the sample farmers in their overall participation in the activities related to WUAs is another question that we looked at across all the three groups of WUAs. Among the three groups of WUAs, the percentage of respondents who expressed that they either very actively or actively participated are found to be very high in lift irrigation system followed by canal irrigation system. For instance, over 93 per cent of sample respondents coming under lift irrigation system expressed that they have either actively or very actively participated in all activities related to WUAs. The same comes to about 75 per cent for the canal irrigation system, but it is only about 63 percent for Pani Panchayat. There are two main reasons as to why the participation is relatively higher among the farmers belonging to lift irrigation system. First, the farmers belonging to lift irrigation societies have invested substantial amount of money in establishing it and therefore, their participation is found to be higher. Second, most of the farmers have been cultivating high value crops under lift irrigation system and therefore, their participation in WUAs is always necessary to sustain the crop cultivation as well as the organisation.

Land Holding Size

Size of land holding is one factor which principally decides the income generating capacity of the households in rural areas. Therefore, along with personal characteristics of the sample respondents, an attempt is made to study the land holding pattern of the households selected from different WUAs. Here, apart from looking at the actual land holdings of the households, details about the area leased-in and leased-out are also considered for analysis.

Table 3.5 presents the details of area operated by sample households selected from three different groups of WUAs. It is surprising to see that the average operated area of the household is almost same across all the three groups (varies from 2.43 ha to 2.75 ha) of the users associations.

Table 3.3: Basic Characteristics of Sample Households-Lift Irrigation System

Particulars	*Farm Size*			
	Small	*Medium*	*Large*	*Overall*
Number of respondents reported as head of household	17 (18)	12 (15)	12 (12)	41 (45)
Age of the head of household (years)	54.22	55.87	58.75	55.98
Education of the respondent (code)	2.72	2.40	2.83	2.64
Farming experience (years)	25.33	32.00	33.00	29.60
Average size of family	6.28	9.00	10.25	8.24
Distance from nearest town/ city (km)	6.33	6.33	6.33	6.33
Position in the Village (Percentage to total sample)				
Head of households	94.44	80.00	100.00	91.11
Village leader	22.22	6.67	41.67	22.22
Position in Water User Association				
Non-member	0	0	0	0
Member	13	12	7	32
Managing committee member	3	2	1	6
Chairman	0	1	3	4
Vice-chairman	0	0	1	1
Secretary	1	0	0	1
Staff	1	0	0	1
Others	0	0	0	0
Participation in activities/ decision-making of WUA (Percentage)				
Very active	33.3	26.7	50.0	35.6
Active	61.1	66.7	41.7	57.8
Passive	5.6	6.7	8.3	6.7
None	0.0	0.0	0.0	0.0

Note and *Source*: Same as in Table 2.1.

Table 3.4: Basic Characteristics of Sample Households-Pani Panchayat

Particulars	*Farm Size*			
	Small	*Medium*	*Large*	*Overall*
Number of respondents reported as head of household	12 (12)	9 (10)	7 (08)	28 (30)
Age of the head of household (years)	53.42	53.90	51.38	53.03
Education of the respondent (code)	2.50	2.60	3.13	2.70
Farming experience (years)	30.83	30.90	22.88	28.73
Average size of family	5.92	7.10	7.25	6.67
Distance from nearest town/city (km)	18.00	18.50	18.00	18.17
Position in the Village (Percentage to total sample)				
Head of households	100.00	90.00	87.50	93.33
Village leader	16.67	10.00	0.00	10.00
Position in Water User Association				
Non-member	0	0	0	0
Member	11	10	8	29
Managing committee member	1	0	0	1
Chairman	0	0	0	0
Vice-chairman	0	0	0	0
Secretary	0	0	0	0
Staff	0	0	0	0
Others	0	0	0	0
Participation in activities/ decision-making of WUA (Percentage)				
Very active	16.7	10.0	25.0	16.7
Active	50.0	50.0	37.5	46.7
Passive	33.3	40.0	37.5	36.7
None	0.0	0.0	0.0	0.0

Note and *Source*: Same as in Table 2.1.

While the average operated area comes to 2.43 ha for the farmers selected from canal WUAs, the same comes to 2.75 ha and 2.62 ha respectively for the farmers selected from lift irrigation and Pani Panchayat WUAs. A little bit prevalence of leased-in and leased-out activities is observed among the members of all the three groups of WUAs. On the whole, though the sample farmers have been selected randomly from different WUAs, our survey data shows that the land size across all the households is almost the same.

Sources of Irrigation

Similar to land size of the households, the availability of irrigation including the source of irrigation is an important factor in determining the output of crop as well as income generating capacity of the farmers' households.[1] Therefore, it is relevant to study the availability of irrigation in different categories of households selected from three different types of WUAs.

It can be seen from Table 3.6 that the availability of irrigation is very high among the sample households selected from different WUAs as compared to the state's average. The percentage of irrigated area to cropped area comes to only 17 per cent for the state (2004-05), but the same comes to over 80 per cent for the sample households selected for the present study. Since the sample farmers have been selected only from the irrigated region and that too from the farmers households registered with WUAs, the percentage of irrigated area is found to be higher. However, as expected, the level of irrigation is not the same across different farm size categories and also between the households coming under different types of WUAs. The level of irrigation of the farmers selected from the lift irrigation water users group is 100 per cent, whereas the same is only 45 per cent for those farmers selected from the group of Pani Panchayat. Since water scarcity is always higher in Pani Panchayat areas and also the objectives of Pani Panchayat are totally different from other WUAs, the percentage of irrigated area is very less among the group of farmers selected from Pani Panchayat areas.

The level of irrigation across different farm size groups is also not the same, except the farmers selected from lift irrigation farmers. While the level of irrigation comes to 100 per cent for all the three size categories of farmers selected from lift irrigation group, the irrigation level is negatively correlated with farm size in canal and Pani Panchayat areas. For instance, the coverage of irrigation comes to 93.52 per cent for small size farmers, but the same is 79.89 per cent for the large farm size

Table 3.5: Average Operated Area of the Sample Households

Area in ha	*Canal Irrigation System*				*Lift Irrigation System*			
	Small	*Medium*	*Large*	*Overall*	*Small*	*Medium*	*Large*	*Over*
Average owned area	1.30	2.46	5.16	2.72	1.08	2.72	4.20	2.4
Average leased in area	0.13	0.00	0.00	0.05	0.09	0.03	0.00	0.0
Average leased out area	0.00	0.05	0.00	0.02	0.00	0.09	0.13	0.0
Average total operated area	1.43	2.41	5.16	2.75	1.16	2.65	4.07	2.4

Area in ha	*Pani Panchayat*				*All*			
	Small	*Medium*	*Large*	*Overall*	*Small*	*Medium*	*Large*	*Ove*
Average owned area	1.00	2.62	5.01	2.61	1.17	2.57	4.84	2.
Average leased in area	0.20	0.00	0.00	0.08	0.13	0.01	0.00	0.
Average leased out area	0.00	0.20	0.00	0.07	0.00	0.09	0.04	0.
Average total operated area	1.20	2.42	5.01	2.62	1.30	2.48	4.80	2.

Source: Field survey data.

Table 3.6: Availability of Irrigation for the Selected Sample Households

Sources of irrigation	*Farm Size*			*Overall*
	Small	*Medium*	*Large*	
	Percentage of Sample Households			
Canal	50.0	50.0	50.0	50.0
Lift System	30.0	30.0	30.0	30.0
Pani Panchayat	20.0	20.0	20.0	20.0
	100.0	100.0	100.0	100.0
Level of Irrigation				
WUA - Canal				
Average irrigated area (ha)	1.30	2.46	5.16	2.72
Percent of NIA to operated area	93.52	82.48	79.89	83.48
WUA - Lift				
Average irrigated area (ha)	1.16	2.65	4.07	2.43
Percent of NIA to operated area	100.00	100.00	100.00	100.00
WUA - Pani Panchayat				
Average irrigated area (ha)	0.66	1.44	1.65	1.18
Percent of NIA to operated area	54.86	59.63	32.92	45.14
All WUA				
Average irrigated area (ha)	1.15	2.08	3.61	2.12
Percent of NIA to operated area	88.15	83.64	75.19	80.42

Note: NIA- net irrigated area.
Source: Field survey data.

selected from the canal irrigated areas. Similarly, the percentage coverage of irrigation comes to 54.86 per cent for the small size farmers, whereas the same comes to only 32.92 per cent for the large size farmers selected from Pani Panchayat areas. This is not an unexpected result because the national level picture also shows a negative relationship between the coverage of irrigation and farm size in India (see, Narayanamoorthy, 2002).

On the whole, except in the case of land size, the above analysis shows considerable differences in the personal characteristics of the sample households selected from different types of WUAs. In the following chapters, let us see how the differences noted in personal characteristics affect the participation as well as the overall performance of WUAs selected for the analysis.

Note

1. The impact of irrigation on productivity of crop has been brought out by a number of studies using both micro and macro-level data in India. For more details readers are suggested to see, Dhawan (1988) and Vaidyanathan, et al., (1994).

4

ROLE AND FUNCTIONS OF WUAs

What is the role of various functionaries and socio-economic groups in the functioning of the Water Users' Associations and to what extent the devolution of power to the WUAs has taken place in the State are two of the main objectives of the book. Therefore, after having analysed the basic characteristics of the sample households, we have focused our analysis to find out the answers to these two questions using the data collected from our sample respondents. Since water plays a crucial role in every aspect of human life, involvement from different stakeholders in decision making process is important for the success of WUAs. Keeping this in view, farmers were asked to indicate the involvement of different groups in terms of their participation in the overall functioning of WUAs. This analysis is expected to throw some light about the involvement of different stakeholders as well as the devolution of powers given to WUAs.

Role of Various Functionaries in WUAs

In order to find out how actively different stakeholders are involved in the functioning of WUAs, the sample farmers were asked indicate their perceptions on 30 different parameters listed in Table 4.1 Since the sample farmers have been selected from three different types of WUAs functioning under canal irrigated area, lift irrigated area and Pani Panchayat area, farmers perceptions about the role of different groups are expected to be different. Therefore, we have analysed the farmers' responses on various parameters separately by each Users Association so as to find out whether any difference exists or not among them. Let us first study the overall responses of the farmers selected from all the three types of WUAs.

Table 4.1 presents the overall responses of the farmers on various socio-economic groups in the functioning of WUAs. It is surprising to see from the overall results that the very active participation is very low from different stakeholders in various activities pertaining to WUAs. Though the active role of Chairman is important for the overall performance of WUAs, only 16 per cent of the respondents have expressed that the Chairman's role is very active. However, as expected, around 30 per cent of the respondents expressed that the head reach farmers have been involved very actively in various activities of WUAs.

However, when we look at the different group of stakeholders individually, we are able to see some kind of differences in the role played by each group of stakeholders in the functioning of WUAs. While the role of Panchayat institutions is found to be almost none, the role of farmers group (all size of farmers) and agricultural labourers is found to be relatively better than others. This is expected because the farmers and the agricultural labourers are directly benefited from WUAs than others and therefore, the role played by these groups is much better than others.

The overall responses that we got from different group of farmers might be different from the responses of the farmers coming under different WUAs. For instance, the responses of the farmers selected from canal command areas might be different from the farmers selected from lift irrigated areas. Since the institutions that control each WUA are different besides source of water supply, the responses of the farmers are expected to be different. Therefore, after having looked at the overall perceptions of the sample farmers, we have looked at each WUA separately.

WUA Role in Canal Command Areas

Let us first look at the responses of the farmers selected from WUAs located in the canal command areas. As can be seen from Table 4.2, the responses of the farmers are somewhat different from the overall responses of the farmers that we have seen above. The very active role played by different groups/functionaries is found to be completely absent from the members of WUAs and other farmers. However, considerable percentage of farmers have expressed that the role played by the office bearers of WUAs is active. Large size farmers, government officials, chairman and secretary of WUAs as well as tail end farmers reported to have played active role in the functioning of WUAs working in the canal command areas. Since active role of the farmers and other officer bearers

Table 4.1: Role of Various Functionaries/Socio-Economic Groups in the Functioning of the WUAs – Results of All WUAs

Particulars	*Percentage of Households Reporting*	*Percentage of Reporting Households*			
		Very Active	*Active*	*Passive*	*None*
1. General body	100.0	8.7	52.0	33.3	6.0
2. Chairman	100.0	16.0	68.7	8.7	6.7
3. Managing committee	100.0	11.3	44.0	34.7	10.0
4. Members	100.0	6.7	42.7	48.7	2.0
5. Non-members	100.0	0.0	0.7	3.3	96.0
6. Secretary	100.0	12.0	63.3	14.0	10.7
7. Other staff	100.0	5.3	39.3	16.7	38.7
8. Any other individual	100.0	0.0	0.7	2.7	96.7
9. Government officials	100.0	0.0	29.3	30.7	40.0
10. Panchayat	100.0	0.0	0.0	0.0	100.0
11. Sarpanch	100.0	0.0	2.7	0.0	97.3
12. Cooperative Credit Society	100.0	0.0	10.7	5.3	84.0
13. Cooperative Marketing Society	100.0	0.0	0.0	4.0	96.0
14. Other local institutions	100.0	19.3	6.0	3.3	71.3
15. Any particular religious group	100.0	0.0	10.7	11.3	78.0
16. Any particular caste group	100.0	0.0	0.0	4.0	96.0
17. Any other specific group	100.0	0.0	2.7	2.7	94.7
18. Women	100.0	0.0	9.3	22.7	68.0
19. Poor	100.0	0.0	22.7	7.3	70.0

20. Middle income	100.0	0.0	30.0	13.3	56.7
21. Upper income	100.0	0.0	18.0	9.3	72.7
22. Large/medium farmers	100.0	2.0	84.0	12.0	2.0
23. Small/ marginal farmers	100.0	2.7	63.3	32.0	2.0
24. Landless	100.0	4.7	23.3	19.3	52.7
25. Labour/wage earners	100.0	13.3	49.3	22.0	15.3
26. Livestock owners	100.0	19.3	50.0	16.0	14.7
27. Tribal	100.0	0.0	2.0	9.3	88.7
28. Upper caste	100.0	0.0	18.7	4.0	77.3
29. Lower caste	100.0	0.0	10.7	7.3	82.0
30. Scheduled Castes	100.0	0.0	9.3	1.3	89.3
31. Head reach farmers	50.0	29.3	64.0	6.7	0.0
32. Tail reach farmers	50.0	8.0	57.3	30.7	4.0
33. Youth	100.0	0.0	5.3	4.7	90.0
34. Any other	100.0	0.0	0.0	0.0	100.0
Average	97.1	4.7	26.2	13.0	56.2

Source: Field survey data.

are essential to get water in time and to solve all other issues pertaining to water management in the canal command areas, the role of these groups is reported to be better than other stakeholders. However, unfortunately, the role of village institutions such as Panchayat, cooperative credit society, caste groups, etc., is reported to be very limited in the functioning of WUAs.

WUA Role in Lift Irrigated Area

The role played by different stakeholders in improving the performance of WUAs established in lift irrigated areas are expected to be much better because of two reasons. First, the capital invested by each member of WUA is relatively higher in lift irrigation than that of

Table 4.2: Role of Various Functionaries/Socio-Economic Groups in the Functioning of the WUA - Canal Irrigation System

Particulars	*Percentage of Households Reporting*	*Percentage of Reporting Households*			
		Very Active	*Active*	*Passive*	*None*
1. General body	100.0	0.0	45.3	54.7	0.0
2. Chairman	100.0	2.7	88.0	9.3	0.0
3. Managing committee	100.0	0.0	34.7	62.7	2.7
4. Members	100.0	0.0	21.3	74.7	4.0
5. Non-members	100.0	0.0	1.3	6.7	92.0
6. Secretary	100.0	1.3	74.7	21.3	2.7
7. Other staff	100.0	0.0	14.7	18.7	66.7
8. Any other individual	100.0	0.0	1.3	4.0	94.7
9. Government officials	100.0	0.0	42.7	41.3	16.0
10. Panchayat	100.0	0.0	0.0	0.0	100.0
11. Sarpanch	100.0	0.0	5.3	0.0	94.7
12. Cooperative Credit Society	100.0	0.0	4.0	9.3	86.7
13. Cooperative Marketing Society	100.0	0.0	0.0	8.0	92.0
14. Other local institutions	100.0	0.0	1.3	4.0	94.7
15. Any particular religious group	100.0	0.0	0.0	0.0	100.0
16. Any particular caste group	100.0	0.0	0.0	5.3	94.7
17. Any other specific group	100.0	0.0	1.3	1.3	97.3
18. Women	100.0	0.0	1.3	6.7	92.0

19. Poor	100.0	0.0	1.3	4.0	94.7
20. Middle income	100.0	0.0	4.0	14.7	81.3
21. Upper income	100.0	0.0	10.7	10.7	78.7
22. Large/medium farmers	100.0	1.3	84.0	10.7	4.0
23. Small/ marginal farmers	100.0	1.3	46.7	48.0	4.0
24. Landless	100.0	9.3	2.7	16.0	72.0
25. Labour/wage earners	100.0	26.7	28.0	20.0	25.3
26. Livestock owners	100.0	18.7	41.3	17.3	22.7
27. Tribal	100.0	0.0	0.0	4.0	96.0
28. Upper caste	100.0	0.0	9.3	6.7	84.0
29. Lower caste	100.0	0.0	4.0	13.3	82.7
30. Scheduled Castes	100.0	0.0	1.3	2.7	96.0
31. Head reach farmers	100.0	29.3	64.0	6.7	0.0
32. Tail reach farmers	100.0	8.0	57.3	30.7	4.0
33. Youth	100.0	0.0	1.3	0.0	98.7
34. Any other	100.0	0.0	0.0	0.0	100.0
Average	100.0	2.9	20.4	15.7	61.0

Source: Field survey data.

the farmers utilising water from canal irrigation. Second, unlike canal irrigation, the direct gravity flow of irrigation water for crop cultivation is absent in lift irrigation and therefore, the role of direct stakeholders (farmers and others) is always necessary to get water in time for crop cultivation.

As we expected, the responses reported in Table 4.3 show that either very active or active role played by the members of WUAs and different income groups is found to be much better than that the responses of the farmers selected from canal command areas. Again, the responses of the farmers show that the role of village institutions in the overall functioning of lift irrigation societies is completely absent, which is very

Table 4.3: Role of Various Functionaries/Socio-Economic Groups in the Functioning of the WUA - Lift Irrigation System

Particulars	*Percentage of Households Reporting*	*Percentage of Reporting Households*			
		Very Active	*Active*	*Passive*	*None*
1. General body	100.0	26.7	71.1	2.2	0.0
2. Chairman	100.0	33.3	64.4	2.2	0.0
3. Managing committee	100.0	24.4	73.3	2.2	0.0
4. Members	100.0	17.8	71.1	11.1	0.0
5. Non-members	100.0	0.0	0.0	0.0	100.0
6. Secretary	100.0	22.2	71.1	6.7	0.0
7. Other staff	100.0	17.8	66.7	11.1	4.4
8. Any other individual	100.0	0.0	0.0	0.0	100.0
9. Government officials	100.0	0.0	0.0	0.0	100.0
10. Panchayat	100.0	0.0	0.0	0.0	100.0
11. Sarpanch	100.0	0.0	0.0	0.0	100.0
12. Cooperative Credit Society	100.0	0.0	28.9	2.2	68.9
13. Cooperative Marketing Society	100.0	0.0	0.0	0.0	100.0
14. Other local institutions	100.0	64.4	17.8	0.0	17.8
15. Any particular religious group	100.0	0.0	35.6	37.8	26.7
16. Any particular caste group	100.0	0.0	0.0	0.0	100.0
17. Any other specific group	100.0	0.0	0.0	0.0	100.0
18. Women	100.0	0.0	26.7	57.8	15.6
19. Poor	100.0	0.0	73.3	8.9	17.8

20. Middle income	100.0	0.0	91.1	4.4	4.4
21. Upper income	100.0	0.0	35.6	2.2	62.2
22. Large/medium farmers	100.0	0.0	95.6	4.4	0.0
23. Small/marginal farmers	100.0	6.7	88.9	4.4	0.0
24. Landless	100.0	0.0	71.1	20.0	8.9
25. Labour/wage earners	100.0	0.0	88.9	11.1	0.0
26. Livestock owners	100.0	26.7	68.9	2.2	2.2
27. Tribal	100.0	0.0	6.7	24.4	68.9
28. Upper caste	100.0	0.0	33.3	2.2	64.4
29. Lower caste	100.0	0.0	28.9	0.0	71.1
30. Scheduled Castes	100.0	0.0	28.9	0.0	71.1
31. Head reach farmers	0.0	0.0	0.0	0.0	0.0
32. Tail reach farmers	0.0	0.0	0.0	0.0	0.0
33. Youth	100.0	0.0	13.3	0.0	86.7
34. Any other	100.0	0.0	0.0	0.0	100.0
Average	100.0	7.1	36.8	6.4	43.9

Source: Field survey data.

unfortunate.

Pani Panchayat is one of the WUAs which is formed with the support of an NGO in order to protect the livelihood of the farmers living in the rainfed areas. Unlike the farmers registered with canal irrigation users group, the direct control of the government is very negligible in regulating the performance of WUAs. Therefore, one would expect relatively better role from different stakeholders in the functioning of WUAs formed in the Pani Panchayat areas. However, against our expectation, the score presented in Table 4.4 shows that the very active role played by the different stakeholders is not much different from other two WUAs. Surprisingly, even the members' role is reported to be very poor in the overall functioning of WUAs. Our deeper inquiry on this

Table 4.4: Role of Various Functionaries/Socio-Economic Groups in the Functioning of the WUA - Pani Panchayat

Particulars	*Percentage of Households Reporting*	*Percentage of Reporting Households*			
		Very Active	*Active*	*Passive*	*None*
1. General body	100.0	3.3	40.0	26.7	30.0
2. Chairman	100.0	23.3	26.7	16.7	33.3
3. Managing committee	100.0	20.0	23.3	13.3	43.3
4. Members	100.0	6.7	53.3	40.0	0.0
5. Non-members	100.0	0.0	0.0	0.0	100.0
6. Secretary	100.0	23.3	23.3	6.7	46.7
7. Other staff	100.0	0.0	60.0	20.0	20.0
8. Any other individual	100.0	0.0	0.0	3.3	96.7
9. Government officials	100.0	0.0	40.0	50.0	10.0
10. Panchayat	100.0	0.0	0.0	0.0	100.0
11. Sarpanch	100.0	0.0	0.0	0.0	100.0
12. Cooperative Credit Society	100.0	0.0	0.0	0.0	100.0
13.Cooperative Marketing Society	100.0	0.0	0.0	0.0	100.0
14.Other local institutions	100.0	0.0	0.0	6.7	93.3
15. Any particular religious group	100.0	0.0	0.0	0.0	100.0
16. Any particular caste group	100.0	0.0	0.0	6.7	93.3
17. Any other specific group	100.0	0.0	10.0	10.0	80.0
18. Women	100.0	0.0	3.3	10.0	86.7
19. Poor	100.0	0.0	0.0	13.3	86.7
20. Middle income	100.0	0.0	3.3	23.3	73.3

21. Upper income	100.0	0.0	10.0	16.7	73.3
22. Large/medium farmers	100.0	6.7	66.7	26.7	0.0
23. Small/ marginal farmers	100.0	0.0	66.7	33.3	0.0
24. Landless	100.0	0.0	3.3	26.7	70.0
25. Labour/wage earners	100.0	0.0	43.3	43.3	13.3
26. Livestock owners	100.0	10.0	43.3	33.3	13.3
27. Tribal	100.0	0.0	0.0	0.0	100.0
28. Upper caste	100.0	0.0	20.0	0.0	80.0
29. Lower caste	100.0	0.0	0.0	3.3	96.7
30. Scheduled Castes	100.0	0.0	0.0	0.0	100.0
31. Head reach farmers	0.0	0.0	0.0	0.0	0.0
32. Tail reach farmers	0.0	0.0	0.0	0.0	0.0
33. Youth	100.0	0.0	3.3	23.3	73.3
34. Any other	100.0	0.0	0.0	0.0	100.0
Average	100.0	2.7	15.9	13.3	62.2

Source: Field survey data.

reveals that the members have lost interest on Pani Panchayat institutions because of inadequate water supply, restrictions for cultivating water-intensive crops and absence of dedicated support from the leader, who was instrumental in forming these societies earlier.[1] Similar to other WUAs, the role of village institutions is also reported to be very inactive in the overall functioning of Pani Panchayat.

Devolution of Powers

Apart from looking at the role played by different stakeholders in the functioning of WUAs, we have also made an attempt to see the changes that have taken place in the devolution of powers given to WUAs after its establishment. Studies have shown that devolution of power is necessary to accomplish the specified duties and objectives of WUAs in an effective manner. Research also tends to suggest that with

Table 4.5: Devolution of Powers – Results of All WUAs

Sr. No.	Decision / Activity	Just Before WUA Formation					At Present				
		Percentage of HH reporting	*Govt.*	*WUA / Farmers Body*	*Joint*	*Others*	*Percentage of HH reporting*	*Govt.*	*WUA / Farmers Body*	*Joint*	*Others*
		Percentage of Households Reporting									
1.	Planning for capital investment in irrigation structures	80	62.5	37.5	0.0	0.0	100.0	7.3	80.0	10.7	2.0
2.	Providing resources for investment	80	62.5	37.5	0.0	0.0	100.0	8.7	65.3	24.0	2.0
3.	Actual capital investment in irrigation structures	80	62.5	37.5	0.0	0.0	100.0	2.7	75.3	21.3	0.7
4.	Assessment of water availability	80	62.5	36.7	0.8	0.0	100.0	49.3	50.0	0.7	0.0
5.	Planning for release of water	80	62.5	37.5	0.0	0.0	100.0	50.7	48.7	0.7	0.0

6.	Actual release of water	80	62.5	37.5	0.0	0.0	100.0	52.0	48.0	0.0	0.0
7.	Distribution of water among farmers	80	62.5	37.5	0.0	0.0	100.0	2.0	97.3	0.7	0.0
8.	Pricing of water received	80	87.5	12.5	0.0	0.0	100.0	74.0	24.7	1.3	0.0
9.	Pricing of water distributed to farmers	80	62.5	37.5	0.0	0.0	100.0	9.3	90.0	0.7	0.0
10.	Collection of dues from farmers	80	62.5	37.5	0.0	0.0	100.0	9.3	82.0	8.7	0.0
11.	Decision on maintenance/ repair requirement	80	62.5	37.5	0.0	0.0	100.0	0.7	98.7	0.7	0.0
12.	Providing resources for maintenance/ repair	80	62.5	37.5	0.0	0.0	100.0	5.3	78.7	16.0	0.0
13.	Implementation of maintenance/ repair	80	62.5	37.5	0.0	0.0	100.0	1.3	98.0	0.7	0.0

14.	Monitoring use of water	80	62.5	37.5	0.0	0.0	100.0	23.3	74.7	2.0	0.0
15.	Stopping misuse/ waste	80	62.5	37.5	0.0	0.0	100.0	1.3	83.3	15.3	0.0
16.	Action on misusers	80	62.5	37.5	0.0	0.0	100.0	2.0	90.0	8.0	0.0
17.	Crops to be grown	80	60.0	1.7	0.0	38.3	100.0	0.0	65.3	2.0	32.7
18.	Others	18	100	0.0	0.0	0.0	22.7	0.0	100.0	0.0	0.0
	Average	73.6	65.8	32.0	0.0	2.1	95.7	16.6	75.0	6.3	2.1

Source: Field survey data.

Table 4.6: Devolution of Powers - Canal Irrigation System

Sr. No.	Decision / Activity	*Just Before WUA Formation*					*At Present*				
		Percentage of HH reporting	*Govt.*	*WUA/ Farmers Body*	*Joint*	*Others*	*Percentage of HH reporting*	*Govt.*	*WUA/ Farmers Body*	*Joint*	*Others*
		Percentage of Households Reporting									
1.	Planning for capital investment in irrigation structures	100.0	100.0	0.0	0.0	0.0	100.0	10.7	76.0	13.3	0.0
2.	Providing resources for investment	100.0	100.0	0.0	0.0	0.0	100.0	17.3	46.7	36.0	0.0
3.	Actual capital investment in irrigation structures	100.0	100.0	0.0	0.0	0.0	100.0	5.3	65.3	29.3	0.0
4.	Assessment of water availability	100.0	100.0	0.0	0.0	0.0	100.0	98.7	1.3	0.0	0.0
5.	Planning for release of water	100.0	100.0	0.0	0.0	0.0	100.0	100	0.0	0.0	0.0

6.	Actual release of water	100.0	100.0	0.0	0.0	0.0	100.0	100	0.0	0.0	0.0
7.	Distribution of water among farmers	100.0	100.0	0.0	0.0	0.0	100.0	2.7	96.0	1.3	0.0
8.	Pricing of water received	100.0	100.0	0.0	0.0	0.0	100.0	84.0	16.0	0.0	0.0
9.	Pricing of water distributed to farmers	100.0	100.0	0.0	0.0	0.0	100.0	0.0	100.0	0.0	0.0
10.	Collection of dues from farmers	100.0	100.0	0.0	0.0	0.0	100.0	0.0	98.7	1.3	0.0
11.	Decision on maintenance/ repair requirement	100.0	100.0	0.0	0.0	0.0	100.0	1.3	97.3	1.3	0.0
12.	Providing resources for maintenance/ repair	100.0	100.0	0.0	0.0	0.0	100.0	10.7	57.3	32.0	0.0
13.	Implementation of maintenance/ repair	100.0	100.0	0.0	0.0	0.0	100.0	2.7	96.0	1.3	0.0

14.	Monitoring use of water	100.0	100.0	0.0	0.0	0.0	100.0	45.3	50.7	4.0	0.0
15.	Stopping misuse/ waste	100.0	100.0	0.0	0.0	0.0	100.0	0.0	73.3	26.7	0.0
16.	Action on misusers	100.0	100.0	0.0	0.0	0.0	100.0	1.3	82.7	16.0	0.0
17.	Crops to be grown	100.0	96.0	2.7	0.0	1.3	100.0	0.0	90.7	4.0	5.3
18.	Others	36.0	100.0	0.0	0.0	0.0	30.7	0.0	100.0	0.0	0.0
	Average	96.4	99.8	0.1	0.0	0.1	96.1	26.7	63.8	9.3	0.3

Source: Field survey data.

Table 4.7: Devolution of Powers - Lift Irrigation System

Sr. No.	Decision / Activity	*Just Before WUA Formation*					*At Present*				
		Percentage of HH reporting	*Govt.*	*WUA / Farmers Body*	*Joint*	*Others*	*Percentage of HH reporting*	*Govt.*	*WUA / Farmers Body*	*Joint*	*Others*
		Percentage of Households Reporting									
1.	Planning for capital investment in irrigation structures	100.0	0.0	100.0	0.0	0.0	100.0	0.0	97.8	2.2	0.0
2.	Providing resources for investment	100.0	0.0	100.0	0.0	0.0	100.0	0.0	100.0	0.0	0.0
3.	Actual capital investment in irrigation structures	100.0	0.0	100.0	0.0	0.0	100.0	0.0	100.0	0.0	0.0
4.	Assessment of water availability	100.0	0.0	97.8	2.2	0.0	100.0	0.0	97.8	2.2	0.0
5.	Planning for release of water	100.0	0.0	100.0	0.0	0.0	100.0	2.2	97.8	0.0	0.0

6.	Actual release of water	100.0	0.0	100.0	0.0	0.0	100.0	0.0	100.0	0.0	0.0
7.	Distribution of water among farmers	100.0	0.0	100.0	0.0	0.0	100.0	2.2	97.8	0.0	0.0
8.	Pricing of water received	100.0	66.7	33.3	0.0	0.0	100.0	62.2	37.8	0.0	0.0
9.	Pricing of water distributed to farmers	100.0	0.0	100.0	0.0	0.0	100.0	0.0	100.0	0.0	0.0
10.	Collection of dues from farmers	100.0	0.0	100.0	0.0	0.0	100.0	0.0	100.0	0.0	0.0
11.	Decision on maintenance/ repair requirement	100.0	0.0	100.0	0.0	0.0	100.0	0.0	100.0	0.0	0.0
12.	Providing resources for maintenance/ repair	100.0	0.0	100.0	0.0	0.0	100.0	0.0	100.0	0.0	0.0
13.	Implementation of maintenance/ repair	100.0	0.0	100.0	0.0	0.0	100.0	0.0	100.0	0.0	0.0

14.	Monitoring use of water	100.0	0.0	100.0	0.0	0.0	100.0	0.0	100.0	0.0	0.0
15.	Stopping misuse/ waste	100.0	0.0	100.0	0.0	0.0	100.0	0.0	100.0	0.0	0.0
16.	Action on misusers	100.0	0.0	100.0	0.0	0.0	100.0	0.0	100.0	0.0	0.0
17.	Crops to be grown	100.0	0.0	0.0	0.0	100	100.0	0.0	0.0	0.0	100
18.	Others (Specify)	0.0	0.0	0.0	0.0	0.0	0.0	0.0	0.0	0.0	0.0
	Average	94.4	3.7	85.1	0.1	5.6	94.4	3.7	84.9	0.2	5.6

Source: Field survey data.

Table 4.8: Devolution of Powers - Pani Panchayat

Sr. No.	Decision / Activity	*Just Before WUA Formation*					*At Present*				
		Percentage of HH reporting	*Govt.*	*WUA / Farmers Body*	*Joint*	*Others*	*Percentage of HH reporting*	*Govt.*	*WUA / Farmers Body*	*Joint*	*Others*
	Percentage of Households Reporting										
1.	Planning for capital investment in irrigation structures	–	–	–	–	–	100.0	10.0	63.3	16.7	10.0
2.	Providing resources for investment	–	–	–	–	–	100.0	0.0	60.0	30.0	10.0
3.	Actual capital investment in irrigation structures	–	–	–	–	–	100.0	0.0	63.3	33.3	3.3
4.	Assessment of water availability	–	–	–	–	–	100.0	0.0	100.0	0.0	0.0
5.	Planning for release of water	–	–	–	–	–	100.0	0.0	96.7	3.3	0.0

6.	Actual release of water	–	–	–	–	–	100.0	10.0	90.0	0.0	0.0
7.	Distribution of water among farmers	–	–	–	–	–	100.0	0.0	100.0	0.0	0.0
8.	Pricing of water received	–	–	–	–	–	100.0	66.7	26.7	6.7	0.0
9.	Pricing of water distributed to farmers	–	–	–	–	–	100.0	46.7	50.0	3.3	0.0
10.	Collection of dues from farmers	–	–	–	–	–	100.0	46.7	13.3	40.0	0.0
11.	Decision on maintenance/ repair requirement	–	–	–	–	–	100.0	0.0	100.0	0.0	0.0
12.	Providing resources for maintenance/ repair	–	–	–	–	–	100.0	0.0	100.0	0.0	0.0
13.	Implementation of maintenance/ repair	–	–	–	–	–	100.0	0.0	100.0	0.0	0.0

14.	Monitoring use of water	–	–	–	–	–	100.0	3.3	96.7	0.0	0.0
15.	Stopping misuse/ waste	–	–	–	–	–	100.0	6.7	83.3	10.0	0.0
16.	Action on misusers	–	–	–	–	–	100.0	6.7	93.3	0.0	0.0
17.	Crops to be grown	–	–	–	–	–	100.0	0.0	100.0	0.0	0.0
18.	Others (Specify)	–	–	–	–	–	36.7	0.0	100.0	0.0	0.0
	Average						96.5	10.9	79.8	8.0	1.3

Source: Field survey data.

less interference from the State agency in the overall management of irrigation water, WUAs would be able to improve not only the irrigation water use efficiency but even the financial recovery as well.[2]

In order to get reasonable idea about the changes in the devolution of powers after the establishment of WUAs, we have asked the farmers to indicate the position prior to the establishment of WUAs and the experience about the present condition. Eighteen different subject questions were asked from farmers covering the major issues such as investment, water management, pricing and financial recovery, repair and maintenance as well as penalties imposed for violating the rules, etc. Let us now look at the farmers' responses about the devolution of powers on different parameters.

Table 4.5 presents the overall farmers' responses about the changes in the devolution of powers before and after the establishment of WUAs. As expected, the sample farmers have expressed that they could notice considerable changes in the devolution of power from the state agency to the users group after establishing WUAs. The control of state agency was reported to be over 60 to 100 per cent in different subject areas before the establishment of WUAs. But, this has totally changed in most cases after the establishment of WUAs. Around 50 per cent of the sample farmers feel that the state control still exists particularly in the case of planning, releasing and distribution of water. Since the control water supply from the reservoirs/dams is with the state agency, farmers expressed that there is no change in the devolution power even after the establishment of WUAs.

Some significant changes in the devolution of power have been noted from the responses of the farmers selected from the canal command areas (see, Table 4.6). Before the formation of WUAs in the canal command areas, the state agency (irrigation department) was controlling almost all the issues pertaining to irrigation water. But this has changed dramatically after their establishment except in the case of assessment and release of water. Since the assessment of water availability and releasing date of water is partly controlled by the natural factors (monsoon, etc), one cannot expect any dramatic change in this even after handing over the control of irrigation water management to farmers group. Considerable changes in the devolution of power from state control to WUAs in aspects such as pricing of water, collection of dues and maintenance have also been reported by the sample farmers. Since MMISF Act enacted by the State government guarantees to provide devolution of power on various aspects of irrigation water management, these changes in the devolution of power were not unexpected. The responses of the farmers selected from the Canal Water Users' group clearly suggest that considerable changes in the devolution of power after the establishment of WUAs. It is expected the devolution of power given to WUAs would help to take appropriate action as and when needed based on the ground level condition. This might also improve the overall efficiency of the irrigation sector in the long run.

Devolution of Power in Lift Irrigation System Group Vs. Canal Irrigation System Group

While WUAs formed under canal areas are fully supported and monitored by the state agency, the lift irrigation societies are formed through farmers own initiatives and resources, under the Act of lift irrigation cooperative societies.[3] Therefore, except in the case of pricing of water, the farmers associated with the lift irrigation water users group enjoy full autonomy in every aspects of irrigation water. Because of this, the question of devolution of power does not arise in the lift irrigation societies formed in Maharashtra. The farmers responses presented in Table 4.7 clearly show not much change in the devolution of power before and after the establishment WUAs.

Unlike the Water Users' Associations established in canal and lift irrigated areas, Pani Panchayat institutions have been established in the rainfed areas, where water scarcity is very prominent. Moreover, Pani Panchayat is established with the support of a non-governmental organisation so as to protect the livelihood of the farmers under certain unusual conditions.[4] Therefore, the devolution of power has been part of the Pani Panchayat since its inception. Which is also reflected from the responses provided by the sample farmers (see, Table 4.8). Except in the case of pricing and distribution of water, the control of state agency is very negligible in all other cases. Since electricity supply is under the control of state agency and it is essential for supply of water in time for crop cultivation, over 45 to 66 per cent of farmers belonging to Pani Panchayat have expressed that the devolution of power is still with the state agency.

On the whole, the analysis on the role of different functionaries in the overall functioning of WUAs shows that the role played by the direct stakeholders (farmers, wage labourers and office bearers of WUAs) of water is relatively more active than other stakeholders of the village community. The role of village institutions such as Panchayat, Sarpanch, cooperative credit societies, etc, is reported to be very negligible in the overall functioning of WUAs. On the aspect of devolution of power, majority of the sample farmers have expressed that they could notice significant changes in devolution of power from state agency to WUAs after its establishment. However, majority of the farmers reported that they could not see any change in the devolution of power with regard to planning and release of canal water, after the formation of WUAs.

Notes

1. Quite a few studies have looked at the performance of the Pani Panchayats in Maharashtra since the inception of this institution. Though it has helped initially the poor farmers located in the rainfed areas of the state, the performance has been deteriorating over the years because of poor support from the State government and other reasons. The reasons for the poor performance of Pani Panchayat are vividly brought out by Deshpande and Reddy (1990); Thakur and Pattnaik, (2002) and Keremanc, et al., (2006).
2. More details on this issue can be seen from Vermillion, (1997).
3. In order to form a lift irrigation schemes in Maharashtra, at least five members having cultivable land of 15 hectares should join for application. The project will have to be sanctioned by NABARD, if financial assistance is to be availed by the society. If water is to be lifted from river then permission of the State's irrigation department will have to be acquired. In case of water is lifted from well and other groundwater sources then certificate from the district Groundwater Survey Officer is necessary (http://www.mah.nic.in/sahakaar/Nirmann.htm).
4. An overview about the formation of Pani Panchayat including its rules and regulations are elaborately explained in Keremane, et al., (2006).

5

IMPACT OF WUAs ON FARM ECONOMY

One of the major objectives of establishing Water Users' Association is to improve the performance of agricultural economy. According to Brewer et al., (1999), the transfer of management from State to farmers group will provide three major beneficial results. First, it will increase crop production because farmers can collectively take more effective water distribution. Second, farmers likely to take better care of their system than government officials, which will improve the sustainability of the system. Third, management transfer will result both in reduction of staff and their government costs for irrigation management, thus reducing pressure on the exchequer. Studies carried out in different parts of the world including India also suggest that the formation of WUAs helps the farmers to achieve higher productivity besides improving the overall efficiency of the irrigation sector (Vermillion, 1997). Therefore, in this Chapter, we have made an attempt to find out what impact the formation of these Associations have made specially on (a) cropping pattern, (b) input use, and (c) productivity of crops. Similar to previous chapter, the impact of WUAs on the farm economy is studied separately by each WUA because of varied use of source of irrigation water.

Cropping Pattern

Irrigation water availability is one of the principal factors which decides the cropping pattern of a region. Water-intensive and high remunerative crops are generally cultivated in the irrigated regions, whereas low value crops are cultivated in the rainfed regions. Several studies have also corroborated that the introduction of irrigation allows the farmers to shift the crop pattern from low value to high value crops (see, Dhawan, 1988; Vaidyanathan, 1994; Vaidyanathan, et al., 1994). We have seen in the previous chapter that the water availability for farmers has increased considerably after the formation of WUAs. Let us now look at whether the formation of WUAs has made any impact on the cropping pattern in the selected regions.

Table 5.1 presents the overall cropping pattern including the area irrigated before and after the formation of WUAs. As expected, we could observe changes in the cropping pattern as well as in area under irrigation after the formation of WUAs. While the area allotted for kharif season has marginally declined, the same has increased marginally in rabi and summer seasons after the formation of WUAs. Importantly, area allotted for foodgrain crops has declined in both kharif and rabi seasons. The area under foodgrain crops has declined by almost 35 per cent in kharif season and around six per cent in rabi season. Since foodgrain crops do not give much profit, farmers may have increased the area under non-foodgrain crops such as oilseeds and vegetables. Interestingly, area allotted for summer crops has increased by about four times (from 0.11 ha/household to 0.44/ha) after the establishment of WUAs.

Area under irrigated crop has also increased after the formation of WUAs. The percent of irrigated crop area was about 73 per cent before the formation of WUAs, but it increased to nearly 92 after the establishment of WUAs. In fact, we could notice significant increase in the percentage of irrigated area to cropped area in summer season as compared to other two seasons. The increase in irrigated area is about 313 per cent in summer season, whereas the same is only about 24 and 17 per cent respectively in kharif and rabi seasons. This increase in irrigated area is achieved because of increased accountability of irrigation water use after the establishment of WUAs.

Though we could notice a substantial change in the overall crop pattern as well as increase in area under irrigation after the formation of WUAs, this might vary among different farmers selected from various users Associations. Therefore, we try to find out the impact of WUAs on cropping pattern and area under irrigation separately in each WUA. Table 5.2 provides the details of change in cropping pattern and irrigated area for the farmers selected from canal irrigated areas. Against our expectation, the cropped area among the farmers selected from canal irrigated areas has declined marginally mainly because of reduction in area allotted for foodgrain crops. However, the area allotted for better value crops such as vegetables, etc., has increased in all the three seasons, including summer season. The share of irrigated cropped area has also

Table 5.1: Change in Cropping Pattern and Area Irrigated – Results of All WUAs

Season/Name of Crop	At the time of WUA Formation		At Present		Percentage Change	
	Average Area under the Crop (ha/hh)	Percent of Area Irrigated	Average Area under the Crop (ha/hh)	Percent of Area Irrigated under the Crop	Average Area under the Crop (ha/hh)	Area Irrigated under the Crop
Kharif						
Foodgrains	1.16	45.71	0.76	72.39	-34.68	3.45
Oilseeds	0.34	86.04	0.31	97.82	-11.03	1.15
Vegetables	0.11	88.26	0.27	95.98	133.45	153.86
Others	0.06	100.00	0.13	100.00	120.53	120.53
Total	1.68	58.71	1.46	84.42	-13.12	24.92
Rabi						
Foodgrains	1.05	82.30	0.99	93.65	-5.82	7.16
Oilseeds	0.01	100.00	0.01	100.00	-40.00	-40.00
Vegetables	0.22	99.41	0.36	99.63	59.10	59.46
Others	0.04	100.00	0.06	100.00	32.16	32.16
Total	1.33	85.95	1.41	95.46	6.06	17.79

Summer						
Foodgrains	0.02	100.00	0.14	100.00	481.11	481.11
Oilseeds	0.07	100.00	0.20	100.00	189.42	189.42
Vegetables	0.01	100.00	0.03	100.00	225.00	225.00
Others	0.00	-	0.07	100.00	-	-
Total	0.11	100.00	0.44	100.00	313.13	313.13
Perennial						
Fruit crops	0.02	100.00	0.02	100.00	0.00	0.00
Sugarcane	0.17	100.00	0.17	100.00	0.00	0.00
Vegetables	0.01	100.00	0.01	100.00	0.00	0.00
Others	0.04	100.00	0.04	100.00	0.00	0.00
Total	0.23	100.00	0.23	100.00	0.00	0.00
All Total	3.35	73.71	3.54	91.77	5.80	31.72

Note: ha/hh – hectare per household.
Source: Field survey data.

Table 5.2: Change in Cropping Pattern and Area Irrigated - Canal Irrigation System

Season / Name of Crop	At the time of WUA Formation		At Present		Percentage Change	
	Average Area under the Crop (ha / hh)	Percentage of Area Irrigated	Average Area under the Crop (ha / hh)	Percentage of Area Irrigated under the Crop	Average Area under the Crop (ha / hh)	Area Irrigated under the Crop
Kharif						
Foodgrains	1.24	45.74	1.07	83.08	-13.78	56.62
Oilseeds	0.19	91.67	0.18	97.01	-6.94	-1.52
Vegetables	0.04	100.00	0.06	100.00	46.05	46.05
Others	0.00	-	0.01	100.00	-	-
Total	1.48	53.29	1.32	85.84	-10.87	43.58
Rabi						
Foodgrains	1.25	92.09	1.15	99.31	-7.53	-0.29
Oilseeds	0.03	100.00	0.02	100.00	-40.00	-40.00
Vegetables	0.33	100.00	0.36	100.00	7.23	7.23
Others	0.01	100.00	0.01	100.00	0.00	0.00
Total	1.61	93.88	1.53	99.48	-5.00	0.66

Summer						
Foodgrains	0.05	100.00	0.06	100.00	25.00	25.00
Oilseeds	0.09	100.00	0.14	100.00	68.75	68.75
Vegetables	0.02	100.00	0.03	100.00	62.50	62.50
Others	0.00	-	0.01	-	-	-
Total	0.16	100.00	0.25	100.00	56.78	56.78
Perennial						
Fruit crops	0.03	100.00	0.03	100.00	0.00	0.00
Sugarcane	0.14	100.00	0.14	100.00	0.00	0.00
Vegetables	0.01	100.00	0.01	100.00	0.00	0.00
Others	0.07	100.00	0.07	100.00	0.00	0.00
Total	0.25	100.00	0.25	100.00	0.00	0.00
All Total	3.50	77.45	3.35	94.19	-4.34	16.33

Source: Field survey data.

increased from 77 per cent to over 94 per cent after the formation of WUAs. This increase was possible because of better availability of water for farmers in canal command areas after the creation of WUAs.

Positive changes in cropping pattern and irrigated area are also seen among the farmers selected from lift irrigated areas after the creation of WUAs (see, Table 5.3). Farmers have allotted more area for better value crops than foodgrain crops after the establishment of WUAs. The area cultivated during summer season has increased from 0.09 ha/household to 0.86 ha/household. Though the performance of Pani Panchayat has been deteriorating over the years due to various reasons, both cropping pattern and area under irrigation has increased significantly among the farmers selected from Pani Panchayat areas (see, Table 5.4). In fact, the area under irrigation was only 1.21 per cent of the cropped area before the creation of Pani Panchayat, but now this has increased to over 71 per cent.

On the whole, though there are variations in the level of changes that occurred in cropping pattern and area under irrigation among the three types of WUAs selected for the analysis, we could notice considerable changes in cropping pattern as well as improvement in irrigated area after the creation of WUAs in all the three areas. Better participation by farmers in irrigation water management activities and accountability in the use of water after the creation of WUAs are reported to be the main reasons for these positive changes.

Input Use

Studies have shown that the formation of WUAs also helps the farmers to increase the adoption of various inputs in crop cultivation because of reliable water supply as well as changes in cropping pattern from low value to high value crops. Therefore, in our study, we have tried to find out what changes the formation of WUAs have brought in the adoption of various inputs in crop cultivation. Here, instead of assessing the quantity of input use, we have tried to get the qualitative information (in terms of increase or decrease) about the adoption of various inputs after the creation of WUAs.

Table 5.5 presents the overall responses of the farmers about the change in various inputs use after the creation of WUAs. The data presented clearly shows significant positive changes in the adoption of various inputs. Over 24 to 37 per cent of farmers have expressed that they could increase HYV/improved seed and fertilisers to a large extent after the establishment of WUAs. Since local seed does not provide higher

Table 5.3: Change in Cropping Pattern and Area Irrigated – Lift Irrigation System

Season/Name of Crop	*At the time of WUA Formation*		*At Present*		*Percentage Change*	
	Average Area under the Crop (ha/hh)	*Percentage of Area Irrigated*	*Average Area under the Crop (ha/hh)*	*Percentage of Area Irrigated under the Crop*	*Average Area under the Crop (ha/hh)*	*Area Irrigated under the Crop*
Kharif						
Foodgrains	0.94	88.14	0.19	100.00	–79.68	–76.94
Oilseeds	0.74	92.21	0.56	100.00	–23.94	–17.52
Vegetables	0.27	100.00	0.57	100.00	115.33	115.33
Others	0.19	100.00	0.20	100.00	5.31	5.31
Total	2.13	92.08	1.53	100.00	–28.43	–22.28
Rabi						
Foodgrains	0.95	100.00	0.58	100.00	–38.39	–38.39
Oilseeds	0.00	0.00	0.00	0.00	–	–
Vegetables	0.19	100.00	0.48	100.00	152.86	152.86
Others	0.14	100.00	0.16	100.00	16.40	16.40
Total	1.28	100.00	1.23	100.00	–3.97	–3.97
Summer						
Foodgrains	0.00	0.00	0.35	100.00	0.00	0.00

Oilseeds	0.09	100.00	0.34	100.00	287.50	287.50
Vegetables	0.00	0.00	0.01	100.00	–	–
Others	0.00	–	0.16	100.00	–	–
Total	0.09	100.00	0.86	100.00	821.43	821.43
Perennial						
Fruit crops	0.01	100.00	0.01	100.00	0.00	0.00
Sugarcane	0.33	100.00	0.33	100.00	0.00	0.00
Vegetables	0.00	0.00	0.00	0.00	–	–
Others	0.02	100.00	0.02	100.00	0.00	0.00
Total	0.36	100.00	0.36	100.00	0.00	0.00
All Total	3.86	95.62	3.97	100.00	2.83	7.54

Source: Field survey data.

Table 5.4: Change in Cropping Pattern and Area Irrigated - Pani Panchayat

Season/Name of Crop	*At the time of WUA Formation*		*At Present*		*Percentage Change*	
	Average Area under the Crop (ha/hh)	*Percentage of Area Irrigated*	*Average Area under the Crop (ha/hh)*	*Percentage of Area Irrigated under the Crop*	*Average Area under the Crop (ha/hh)*	*Area Irrigated under the Crop*
Kharif						
Foodgrains	1.31	0.00	0.84	28.69	–35.97	0.00
Oilseeds	0.13	10.53	0.24	91.55	86.84	1525.0
Vegetables	0.07	0.00	0.32	83.16	375.0	0.00
Others	0.00	0.00	0.32	100.00	–	–
Total	1.50	0.89	1.71	60.82	14.00	7700.0
Rabi						
Foodgrains	0.69	1.92	1.17	74.93	68.75	6475.0
Oilseeds	0.00	0.00	0.00	0.00	0.00	0.00
Vegetables	0.01	0.00	0.17	96.15	2500.0	–
Others	0.00	0.00	0.04	100.00	–	–
Total	0.70	1.90	1.38	78.26	97.14	8000.0
Summer						
Foodgrains	0.00	0.00	0.03	100.00	–	–

Oilseeds	0.00	0.00	0.13	100.00	–	–
Vegetables	0.00	0.00	0.08	100.00	–	–
Others	0.00	0.00	0.07	100.00	–	–
Total	0.00	0.00	0.30	100.00	–	–
Perennial						
Fruit crops	0.00	0.00	0.00	0.00	–	–
Sugarcane	0.00	0.00	0.00	0.00	–	–
Vegetables	0.00	0.00	0.00	0.00	–	–
Others	0.00	0.00	0.00	0.00	–	–
Total	0.00	0.00	0.00	0.00	–	–
All Total	2.20	1.21	3.39	71.36	53.94	8962.5

Source: Field survey data.

Table 5.5: Change in Input Use Since the Establishment of WUA – Results of All WUAs

Inputs	*Percentage of Households Reporting*	*Large Increase*	*Increase*	*No Change*	*Decrease*	*Large Decrease*
		Percentage of Reporting Households				
Seed local	100.0	2.0	1.3	0.0	61.3	35.3
Seed HYV	100.0	27.3	60.0	9.3	3.3	0.0
Seed improved	100.0	37.3	60.7	1.3	0.7	0.0
Fertilizer	100.0	24.0	73.3	1.3	1.3	0.0
Pesticides	100.0	15.3	81.3	3.3	0.0	0.0
Farm Yard Manure	100.0	10.0	29.3	31.3	29.3	0.0
Bullock labour	100.0	0.7	0.0	31.3	68.0	0.0
Machine labour	100.0	10.0	74.7	14.7	0.7	0.0
Family labour	100.0	6.0	52.0	18.0	21.3	2.7
Hired labour	100.0	4.0	82.7	4.7	8.7	0.0
Cost of irrigation	100.0	10.0	88.7	1.3	0.0	0.0
Other costs (Transportation)	96.0	10.7	84.0	1.3	0.0	0.0

Source: Field survey data.

Table 5.6: Change in Input Use Since the Establishment of WUA – Canal Irrigation System

Inputs	*Percentage of Households Reporting*	*Large Increase*	*Increase*	*No Change*	*Decrease*	*Large Decrease*
		Percentage of Reporting Households				
Seed local	100.0	0.0	1.3	0.0	76.0	22.7
Seed HYV	100.0	13.3	66.7	13.3	6.7	0.0
Seed improved	100.0	33.3	64.0	1.3	1.3	0.0
Fertilizer	100.0	26.7	68.0	2.7	2.7	0.0
Pesticides	100.0	12.0	82.7	5.3	0.0	0.0
Farm Yard Manure	100.0	2.7	26.7	45.3	25.3	0.0
Bullock labour	100.0	0.0	0.0	44.0	56.0	0.0
Machine labour	100.0	4.0	66.7	28.0	1.3	0.0
Family labour	100.0	5.3	66.7	25.3	2.7	0.0
Hired labour	100.0	1.3	86.7	5.3	6.7	0.0
Cost of irrigation	100.0	6.7	92.0	1.3	0.0	0.0
Other costs (Transportation)	93.3	2.9	94.3	2.9	0.0	0.0

Source: Field survey data.

Table 5.7: Change in Input Use Since the Establishment of WUA - Lift Irrigation System

Inputs	*Percentage of Households Reporting*	*Large Increase*	*Increase*	*No Change*	*Decrease*	*Large Decrease*
		Percentage of Reporting Households				
Seed local	100.0	0.0	0.0	0.0	35.6	64.4
Seed HYV	100.0	37.8	62.2	0.0	0.0	0.0
Seed improved	100.0	37.8	62.2	0.0	0.0	0.0
Fertilizer	100.0	26.7	73.3	0.0	0.0	0.0
Pesticides	100.0	26.7	73.3	0.0	0.0	0.0
Farm Yard Manure	100.0	28.9	42.2	4.4	24.4	0.0
Bullock labour	100.0	2.2	0.0	15.6	82.2	0.0
Machine labour	100.0	26.7	73.3	0.0	0.0	0.0
Family labour	100.0	0.0	11.1	13.3	66.7	8.9
Hired labour	100.0	6.7	71.1	4.4	17.8	0.0
Cost of irrigation	100.0	8.9	88.9	2.2	0.0	0.0
Other costs (Transportation)	97.8	22.7	77.3	0.0	0.0	0.0

Source: Field survey data.

Table 5.8: Change in Input Use Since the Establishment of WUA - Pani Panchayat

Inputs	*Percentage of Households Reporting*	*Large Increase*	*Increase*	*No Change*	*Decrease*	*Large Decrease*
		Percentage of Reporting Households				
Seed local	100.0	10.0	3.3	0.0	63.3	23.3
Seed HYV	100.0	46.7	40.0	13.3	0.0	0.0
Seed improved	100.0	46.7	50.0	3.3	0.0	0.0
Fertilizer	100.0	13.3	86.7	0.0	0.0	0.0
Pesticides	100.0	6.7	90.0	3.3	0.0	0.0
Farm Yard Manure	100.0	0.0	16.7	36.7	46.7	0.0
Bullock labour	100.0	0.0	0.0	23.3	76.7	0.0
Machine labour	100.0	0.0	96.7	3.3	0.0	0.0
Family labour	100.0	16.7	76.7	6.7	0.0	0.0
Hired labour	100.0	6.7	90.0	3.3	0.0	0.0
Cost of irrigation	100.0	20.0	80.0	0.0	0.0	0.0
Other costs (Transportation)	100.0	13.3	86.7	0.0	0.0	0.0

Source: Field survey data.

yield, over 96 per cent of farmers could reduce the use of the same after the establishment of WUAs. Significant reduction of bullock labour has also been reported by the farmers, because of added use of machineries such as tractors, etc., in the crop cultivation. However, because of change in input use after the creation of WUAs, the cost of cultivation has reported to be increased as expected for the sample farmers.

Similar to the overall results, farmers selected from Canal Water Users' group have also reported that they could increase the use of various inputs after the establishment of WUAs (Table 5.6). Majority of the farmers reported to have increased the adoption of HYV seed, fertilisers, pesticides, etc., in crop cultivation. Because of change in cultivation due to better availability of water, farmers also reported to have significantly reduced local seed, bullock labour, etc. The same kind of trend is also seen in the use of inputs among the farmers selected from Lift Irrigation Water Users' group as well as from Pani Panchayat group (see, Tables 5.7 and 5.8). On the whole, two important points are emerging out from the analysis of input use after the establishment of WUAs. First, farmers are able to increase use of new technological inputs and reduce low quality inputs. Second, the cost of water and others required for cultivating crops have increased after the creation of WUAs, which may put some hardship to small and resourceless farmers.

Change in Crop Yield on WUA Establishment

What happened to the productivity of various crops cultivated by the farmers after the establishment of WUAs? Many earlier studies have proved that their formation helps increasing the productivity of crops, because they are able to provide better and controlled water supply to the farmers (see, Vermillion, 1997; Brewar, et al., 1999). We have also seen earlier in our study that the formation of These Association have helped the farmers to change the cropping pattern from low value to high value crops and increase the use of new technological inputs. Therefore, one would unsurprisingly expect increase in productivity of different crops after the establishment of the Users Association in our study area.

As in the case of input use, instead of collecting quantitative data, farmers were asked to give qualitative information about the changes that they realised in productivity of crops under five different categories namely large increase, increase, no change, decrease and large decrease. The overall responses of the farmers about the change in productivity of different crops are presented in Table 5.9. As expected, considerable percentage of farmers have expressed that they could realise either large

Table 5.9: Change in Crop Yield Since the Establishment of WUA – Results of All WUAs

Name of Crop	*Percentage of Households Reporting*	*Large Increase*	*Increase*	*No Change*	*Decrease*	*Large Decrease*
		Percentage of Reporting Households				
Irrigated:						
HYV Bajra	78.0	15.4	66.7	16.2	1.7	0.0
Groundnut	52.7	10.1	67.1	8.9	13.9	0.0
Wheat	78.0	12.8	76.9	7.7	2.6	0.0
Gram	38.7	3.4	72.4	20.7	3.4	0.0
Jowar (Fodder)	14.7	0.0	59.1	40.9	0.0	0.0
HYV Jowar	49.3	10.8	67.6	18.9	2.7	0.0
Sugarcane	30.0	11.1	66.7	6.7	15.6	0.0
Maize	28.0	0.0	28.6	71.4	0.0	0.0
Tur	6.0	11.1	66.7	22.2	0.0	0.0
Moog	2.0	0.0	33.3	66.7	0.0	0.0
Paddy	0.7	0.0	100.0	0.0	0.0	0.0
Lucerne	31.3	0.0	34.0	66.0	0.0	0.0
Kadwal	20.0	0.0	26.7	73.3	0.0	0.0
Onion	64.7	20.6	52.6	9.3	17.5	0.0
Tomato	23.3	57.1	42.9	0.0	0.0	0.0

Beat	11.3	17.6	82.4	0.0	0.0	0.0
Cauliflower	11.3	47.1	52.9	0.0	0.0	0.0
Potato	14.0	57.1	42.9	0.0	0.0	0.0
Carrot	10.7	37.5	62.5	0.0	0.0	0.0
Zendu	11.3	58.8	41.2	0.0	0.0	0.0
Unirrigated						
HYV Bajra	27.3	0.0	48.8	41.5	9.8	0.0
HYV Jowar	7.3	0.0	81.8	18.2	0.0	0.0
Tur	2.0	0.0	0.0	100.0	0.0	0.0
Moog	0.7	0.0	0.0	100.0	0.0	0.0
Paddy	0.7	0.0	0.0	100.0	0.0	0.0

Note: Minor crops have been eliminated from the analysis.
Source: Field survey data

Table 5.10: Change in Crop Yield Since the Establishment of WUA - Canal Irrigation System

Name of Crop	*Percentage of Households Reporting*	*Large Increase*	*Increase*	*No Change*	*Decrease*	*Large Decrease*
		Percentage of Reporting Households				
Irrigated						
HYV Bajra	85.3	17.2	68.8	10.9	3.1	0.0
Groundnut	40.0	3.3	46.7	13.3	36.7	0.0
Wheat	94.7	8.5	81.7	5.6	4.2	0.0
Gram	60.0	0.0	68.9	26.7	4.4	0.0
Jowar (Fodder)	1.3	0.0	100.0	0.0	0.0	0.0
HYV Jowar	42.7	0.0	90.6	3.1	6.3	0.0
Sugarcane	32.0	12.5	50.0	8.3	29.2	0.0
Maize	46.7	0.0	14.3	85.7	0.0	0.0
Tur	12.0	11.1	66.7	22.2	0.0	0.0
Lucerne	36.0	0.0	7.4	92.6	0.0	0.0
Kadwal	37.3	0.0	25.0	75.0	0.0	0.0
Onion	76.0	1.8	54.4	14.0	29.8	0.0
Pomegranate	8.0	33.3	66.7	0.0	0.0	0.0
Unirrigated						
HYV Bajra	25.3	0.0	26.3	52.6	21.1	0.0
HYV Jowar	1.3	0.0	100.0	0.0	0.0	0.0
Tur	2.7	0.0	0.0	100.0	0.0	0.0

Note and *Source*: Same as in Table 5.9.

Table 5.11: Change in Crop Yield Since the Establishment of WUA - Lift Irrigation System

Name of Crop	*Percentage of Households Reporting*	*Large Increase*	*Increase*	*No Change*	*Decrease*	*Large Decrease*
		As Percentage of Reporting Households				
Irrigated						
HYV Bajra	93.3	7.1	64.3	28.6	0.0	0.0
Groundnut	75.6	5.9	85.3	8.8	0.0	0.0
Wheat	51.1	8.7	69.6	21.7	0.0	0.0
Jowar Fodder	46.7	0.0	57.1	42.9	0.0	0.0
HYV Jowar	55.6	4.0	44.0	52.0	0.0	0.0
Sugarcane	46.7	9.5	85.7	4.8	0.0	0.0
Maize	13.3	0.0	100.0	0.0	0.0	0.0
Lucerne	44.4	0.0	70.0	30.0	0.0	0.0
Onion	68.9	38.7	58.1	3.2	0.0	0.0
Tomato	42.2	52.6	47.4	0.0	0.0	0.0
Beat	37.8	17.6	82.4	0.0	0.0	0.0
Cauliflower	37.8	47.1	52.9	0.0	0.0	0.0
Potato	46.7	57.1	42.9	0.0	0.0	0.0
Carrot	35.6	37.5	62.5	0.0	0.0	0.0

Sapota	4.4	50.0	50.0	0.0	0.0	0.0
Cabbage	8.9	0.0	25.0	0.0	75.0	0.0
Sunflower	26.7	0.0	83.3	16.7	0.0	0.0
Custard Apple	15.6	14.3	85.7	0.0	0.0	0.0
Unirrigated						
HYV Bajra	0.0	0.0	0.0	0.0	0.0	0.0
HYV Jowar	0.0	0.0	0.0	0.0	0.0	0.0
Tur	0.0	0.0	0.0	0.0	0.0	0.0

Note and *Source*: Same as in Table 5.9.

Table 5.12: Change in Crop Yield Since the Establishment of WUA - Pani Panchayat

Name of Crop	*Percentage of Households Reporting*	*Large Increase*	*Increase*	*No Change*	*Decrease*	*Large Decrease*
		As Percentage of Reporting Households				
Irrigated						
HYV Bajra	36.7	36.4	63.6	0.0	0.0	0.0
Groundnut	50.0	33.3	66.7	0.0	0.0	0.0
Wheat	76.7	30.4	69.6	0.0	0.0	0.0
Gram	43.3	15.4	84.6	0.0	0.0	0.0
Jowar (Fodder)	0.0	0.0	0.0	0.0	0.0	0.0
HYV Jowar	56.7	41.2	58.8	0.0	0.0	0.0
Onion	30.0	77.8	22.2	0.0	0.0	0.0
Tomato	50.0	66.7	33.3	0.0	0.0	0.0
Chilly	10.0	100.0	0.0	0.0	0.0	0.0
Vegetables	16.7	40.0	60.0	0.0	0.0	0.0
Flower (Marigold)	53.3	56.3	43.8	0.0	0.0	0.0
Custard Apple	6.7	50.0	50.0	0.0	0.0	0.0
Unirrigated						
HYV Bajra	73.3	0.0	68.2	31.8	0.0	0.0
HYV Jowar	33.3	0.0	80.0	20.0	0.0	0.0
Hulaga	10.0	0.0	33.3	66.7	0.0	0.0
Matkee	30.0	0.0	22.2	77.8	0.0	0.0
Pavata	10.0	0.0	66.7	33.3	0.0	0.0

Note and *Source*: Same as in Table 5.9.

increase or increase in productivity of major crops after the establishment of WUAs. However, in the case of HYV jowar, maize and moong, a considerable percentage farmers have reported they could not increase the productivity of crops owing to various reasons. Though the response of the farmers about productivity increase varies across crops, the overall trend is that the farmers are able to increase the productivity of major crops after the establishment of WUAs. This is as expected Since the establishment of WUAs has allowed the farmers to use higher quantity of yield increasing technological inputs, thus enabling the farmers to get higher productivity from different crops.

Considerable increase in productivity of major crops has been reported by majority of the farmers selected from different WUAs (see, Tables 5.10, 5.11 and 5.12). About 30 per cent of farmers selected from the Canal Water Users' group reported that the productivity of crops (especially in groundnut, sugar cane and onion) seems to have declined after establishment of WUAs. This reduction in productivity is not mainly because of supply of water but due to diseases and other problems. However, among the three types of WUAs, none of the farmers selected from lift irrigation users' group and Pani Panchayat reported any decrease in productivity of crops. As seen earlier, most of the farmers have increased the adoption of technological inputs after the creation of WUAs and therefore, it must have helped the farmers to get higher yield from different crops.

In total, the analysis on the impact of WUAs on the farm economy shows that the farmers have been able to increase area under irrigated crop, change the crop pattern towards better value crops, and increase the adoption of various yield increasing inputs as well as productivity of crops after the formation of WUAs. However, one should not forget the fact that these positive changes have also increased the cost of cultivation for farmers in all the three areas where the survey has been conducted for the present study.

6

IMPACT OF WUAs ON GENERAL ECONOMY

In the previous chapter, we have studied the impact of WUAs specifically on cropping pattern, input use and productivity of crops. Besides making direct impact on the farming community, the creation of WUAs is also expected to contribute to the whole economy of the area, though indirectly, because of intensification of agriculture. Increased water availability would increase the land use intensity, cropping intensity and productivity of crops, all of which one way or the other create more employment opportunities as well as increase the wage rate for agricultural labourers. Increased availability of water also allows the farmer to diversify their economic activities from crop husbandry to non-crop husbandry. Therefore, in this chapter, we attempt to find out the impact of the creation of WUAs on the general economy of the selected area. This chapter has been divided into two sections. The impact of WUAs on efficiency, equity, social cohesion and others has been discussed in the first section, while the contributions of WUAs to the economic development of the area has been discussed in the second section.

Impact of WUAs on Efficiency, Equity and Social Cohesion

Needless to mention here that it is very difficult to collect quantitative information from the sample farmers about the efficiency, equity and social cohesion because of various practical reasons. Therefore, in order to know the influence of WUAs on these variable, we have asked the farmers to indicate their view points on several parameters under five pre-decided categories namely highly positive, positive, no impact, negative and highly negative. Table 6.1 presents the overall responses

Table 6.1: Impact of WUA on Efficiency, Equity, Social Cohesion and Others – Results of All WUAs

Parameters	*Percentage of Sample Households Reporting*	*Highly Positive*	*Positive*	*No Impact*	*Negative*	*Highly Negative*
		Percentage of Reporting Households				
A. Efficiency Related						
Timely water availability	100	12.0	53.3	30.7	4.0	0.0
Adequate water availability	100	14.7	81.3	2.7	1.3	0.0
Increase in irrigated area	100	36.7	56.7	6.0	0.7	0.0
Change in cropping pattern in favour of high value crops	100	16.0	74.0	8.7	1.3	0.0
Better mintenance of irrigation structure	100	10.0	64.0	11.3	14.7	0.0
Reduction in the cost of maintenance	100	2.7	56.7	19.3	20.0	1.3
B. Equity Related						
Equitable distribution of water	100	5.3	92.0	2.0	0.7	0.0
Empowerment of farmers to manage irrigation systems	100	3.3	78.7	10.7	7.3	0.0
Volumetric pricing	100	1.3	69.3	27.3	1.3	0.7

All land holders became WUA members with voting rights	100	2.7	56.0	14.7	26.7	0.0
C. Social Cohesion and Empowerment						
Beginning of a sense of ownership by farmers	100	2.7	65.3	14.7	17.3	0.0
Unification of diverse groups in the area	100	4.0	65.3	18.7	12.0	0.0
Freedom to raise resources	100	2.0	92.0	5.3	0.7	0.0
Resolution of disputes and compounding of offenses	100	1.3	50.0	18.0	16.7	14.0
Active involvement of all classes	100	3.3	67.3	14.0	15.3	0.0
D. Others						
More Farm Employment	100	28.0	64.0	7.3	0.7	0.0
Year-round availability of water for irrigation	100	10.7	54.7	28.0	6.7	0.0
Diversification of cropping pattern	100	18.7	66.0	14.7	0.7	0.0
Choice in deciding irrigation timings	100	2.7	53.3	20.7	23.3	0.0
Choice in deciding quantum of water	100	4.7	64.7	7.3	23.3	0.0

Diversified economic activities: dairying	100	27.3	64.0	6.7	2.0	0.0
Diversified economic activities : other non-agriculture	98.7	3.4	45.3	7.4	25.0	18.9
Transfer of power to the WUA	100	3.3	73.3	9.3	14.0	0.0
Transfer of power to WUA to decide water charges	100	6.7	73.3	10.7	9.3	0.0
Others	0.0	0.0	0.0	0.0	0.0	0.0

Source: Field survey data.

of the farmers selected from all the three types of WUAs. The results presented in the table appear to suggest that farmers' responses on different parameters are not the same. On the issue of efficiency, though over 90 per cent of the farmers have indicated that they could see highly positive or positive changes in adequate water availability and increase in irrigated area, about 30 per cent of the farmers have expressed that they have realised 'no impact' on the timely availability of water, after the creation of WUAs. Since the control of water supply is still with the state agency, WUAs could make timely availability of water to all the member farmers. This is particularly true in the case of farmers who are part of the Water Users' Association located at the tail end of canal command area.

As regards the equity related questions, considerable percentage of farmers (around 30 per cent) expressed that the volumetric pricing and voting rights for all WUA members have made no impact or made negative impact on the overall development of the economy in the selected area. This suggests that the farmers are not very much likely to adopt the policy of volumetric pricing, which is one of major conditions imposed by the state to establish WUAs. Has the formation of the users Association made any impact on the social cohesion and empowerment? Though the survey results suggest that these Associations have made considerable impact on sense of ownership, freedom to raise resources and unification of diverse groups in the area, it appears from the responses of the farmers that they have not made significant impact on solving disputes and bringing active involvement of all classes of farmers.

The creation of WUAs appears to have made considerable impact on the farm employment, diversification of cropping pattern and other economic activities (dairying, etc.), as well as transfer of power to WUAs. However, nearly 30 per cent of the farmers expressed that they could see no impact on deciding irrigation timings as well as quantum of water after the creation of WUAs. This is something unexpected because one of the major objectives of creation of WUAs is to decide the irrigation timings and quantum of water supply by farmers themselves. This in a way suggests that the formation of WUAs could not fulfil all its objectives perfectly due to many exogenous and endogenous factors.

The overall responses of the farmers on the efficiency, equity and social cohesion discussed above are almost matching with the responses of the farmers selected from the canal water users' group (see, Table 6.2). Nearly 50 per cent of the farmers belonging to the Canal Water Users' group have felt that WUAs have made no impact or negative

Table 6.2: Impact of WUA on Efficiency, Equity, Social Cohesion and Others - Canal Irrigation System

Parameters	*Percentage of Sample Households Reporting*	*Highly Positive*	*Positive*	*No Impact*	*Negative*	*Highly Negative*
		Percentage of Reporting Households				
A. Efficiency Related						
Timely water availability	100.0	1.3	34.7	56.0	8.0	0.0
Adequate water availability	100.0	10.7	84.0	2.7	2.7	0.0
Increase in irrigated area	100.0	44.0	42.7	12.0	1.3	0.0
Change in cropping pattern in favour of high value crops	100.0	2.7	77.3	17.3	2.7	0.0
Better maintenance of irrigation structure	100.0	2.7	49.3	18.7	29.3	0.0
Reduction in the cost of maintenance	100.0	0.0	40.0	22.7	34.7	2.7
B Equity Related						
Equitable distribution of water	100.0	1.3	93.3	4.0	1.3	0.0
Empowerment of farmers to manage irrigation systems	100.0	1.3	62.7	21.3	14.7	0.0
Volumetric pricing	100.0	1.3	60.0	34.7	2.7	1.3

All land holders became WUA members with voting rights	100.0	0.0	49.3	10.7	40.0	0.0
C. Social Cohesion and Empowerment						
Beginning of a sense of ownership by farmers	100.0	0.0	48.0	17.3	34.7	0.0
Unification of diverse groups in the area	100.0	0.0	46.7	29.3	24.0	0.0
Freedom to raise resources	100.0	0.0	88.0	10.7	1.3	0.0
Resolution of disputes and compounding of offenses	100.0	0.0	46.7	34.7	18.7	0.0
Active involvement of all classes	100.0	0.0	61.3	21.3	17.3	0.0
D. Others						
More Farm Employment	100.0	30.7	53.3	14.7	1.3	0.0
Year-round availability of water for irrigation	100.0	6.7	24.0	56.0	13.3	0.0
Diversification of cropping pattern	100.0	16.0	53.3	29.3	1.3	0.0
Choice in deciding irrigation timings	100.0	0.0	12.0	41.3	46.7	0.0
Choice in deciding quantum of water	100.0	0.0	42.7	10.7	46.7	0.0
Diversified economic activities: dairying	100.0	32.0	64.0	2.7	1.3	0.0

Diversified economic activities : other non-agriculture	100.0	2.7	69.3	1.3	25.3	1.3
Transfer of power to the WUA	100.0	0.0	68.0	13.3	18.7	0.0
Transfer of power to WUA to decide water charges	100.0	6.7	65.3	18.7	9.3	0.0
Others	0.0	0.0	0.0	0.0	0.0	0.0

Source: Field survey data.

impact especially on parameters such as (a) timely availability of water, (b) maintenance of irrigation structure, (c) reduction in cost of maintenance, (d) pricing of irrigation water, (e) unification of diverse groups in the area, (f) resolution of disputes and compounding of offences, (g) round the year availability of irrigation, (h) choice of irrigation times, and (i) choice in deciding quantum of water. Though these problems have been continuing in canal irrigated areas across the country over several decades, one would have expected that the formation of these Associations could have reduced all these problems. Since most of these problems are associated with the supply of water, which is under the control of the state agency, these problems may continue ever in case the supply of water is not made available to farmers in time.

The responses of the farmers selected from the lift irrigation users' group show a much better picture about the overall impact of WUAs on the general economy (see, Table 6.3). Over 90 per cent of the farmers have expressed that they could realise either highly positive or positive impact on all other parameters after the formation of WUAs, except parameters such as (a) resolution of disputes and compounding of offences, and (b) diversified economic activities. Since the farmers from lift irrigation group have invested relatively more than the other group of farmers, it must have motivated the farmers to get involved intensively in all activities related to WUAs. The responses of the sample farmers selected from Pani Panchayat group also show relatively better picture about the impact of WUAs on the general economy of the area, barring a few parameters (see, Table 6.4).

On the whole, what is emerging out from the above is that the farmers by and large could realise many positive changes in the efficiency, equity and social cohesion in the selected area after the creation of WUAs. On an average, over 35 per cent of the sample farmers have expressed that the volumetric pricing does not make any impact on the overall development of the economy. Does this mean that farmers are not willing to accept the volumetric pricing in irrigation water?

Contribution of WUAs to Economic Development

After getting the sense of what farmers think about the efficiency, equity and social cohesion in the selected area after the creation of WUAs, we also try to assess the perception of the farmers about the contribution of WUAs to economic development of the selected area. In order to capture the farmers' perceptions, the sample farmers were asked to indicate their perception about the impact WUAs on over 20

Table 6.3: Impact of WUA on Efficiency, Equity, Social Cohesion and Others - Lift Irrigation System

Parameters	*Percentage of Sample Households Reporting*	*Highly Positive*	*Positive*	*No Impact*	*Negative*	*Highly Negative*
		Percentage of Reporting Households				
A. Efficiency Related						
Timely water availability	100.0	20.0	80.0	0.0	0.0	0.0
Adequate water availability	100.0	20.0	80.0	0.0	0.0	0.0
Increase in irrigated area	100.0	22.2	77.8	0.0	0.0	0.0
Change in cropping pattern in favour of high value crops	100.0	22.2	77.8	0.0	0.0	0.0
Better Maintenance of irrigation structure	100.0	24.4	73.3	2.2	0.0	0.0
Reduction in the cost of maintenance	100.0	8.9	73.3	13.3	4.4	0.0
B. Equity Related						
Equitable distribution of water	100.0	15.6	84.4	0.0	0.0	0.0
Empowerment of farmers to manage irrigation systems	100.0	8.9	91.1	0.0	0.0	0.0

Volumetric pricing	100.0	2.2	97.8	0.0	0.0	0.0
All land holders became WUA members with voting rights	100.0	8.9	86.7	2.2	2.2	0.0
C. Social Cohesion and Empowerment						
Beginning of a sense of ownership by farmers	100.0	2.2	97.8	0.0	0.0	0.0
Unification of diverse groups in the area	100.0	4.4	95.6	0.0	0.0	0.0
Freedom to raise resources	100.0	2.2	97.8	0.0	0.0	0.0
Resolution of disputes and compounding of offenses	100.0	0.0	33.3	2.2	17.8	46.7
Active involvement of all classes	100.0	6.7	82.2	2.2	8.9	0.0
D. Others						
More Farm Employment	100.0	13.3	86.7	0.0	0.0	0.0
Year-round availability of water for irrigation	100.0	11.1	88.9	0.0	0.0	0.0
Diversification of cropping pattern	100.0	11.1	88.9	0.0	0.0	0.0
Choice in deciding irrigation timings	100.0	2.2	97.8	0.0	0.0	0.0

Choice in deciding quantum of water	100.0	6.7	93.3	0.0	0.0	0.0
Diversified economic activities: dairying	100.0	24.4	75.6	0.0	0.0	0.0
Diversified economic activities : other non-agriculture	95.6	0.0	2.3	9.3	25.6	62.8
Transfer of power to the WUA	100.0	6.7	86.7	4.4	2.2	0.0
Transfer of power to WUA to decide water charges	100.0	6.7	86.7	4.4	2.2	0.0
Others	0.0	0.0	0.0	0.0	0.0	0.0

Source: Field survey data.

Table 6.4: Impact of WUA on Efficiency, Equity, Social Cohesion and Others - Pani Panchayat

Parameters	*Percentage of Sample Households Reporting*	*Highly Positive*	*Positive*	*No Impact*	*Negative*	*Highly Negative*
		Percentage of Reporting Households				
A. Efficiency Related						
Timely water availability	100.0	26.7	60.0	13.3	0.0	0.0
Adequate water availability	100.0	16.7	76.7	6.7	0.0	0.0
Increase in irrigated area	100.0	40.0	60.0	0.0	0.0	0.0
Change in cropping pattern in favour of high value crops	100.0	40.0	60.0	0.0	0.0	0.0
Better maintenance of irrigation structure	100.0	6.7	86.7	6.7	0.0	0.0
Reduction in the cost of maintenance	100.0	0.0	73.3	20.0	6.7	0.0
B. Equity Related						
Equitable distribution of water	100.0	0.0	100.0	0.0	0.0	0.0
Empowerment of farmers to manage irrigation systems	100.0	0.0	100.0	0.0	0.0	0.0
Volumetric pricing	100.0	0.0	50.0	50.0	0.0	0.0

All land holders became WUA members with voting rights	100.0	0.0	26.7	43.3	30.0	0.0
C. Social Cohesion and Empowerment						
Beginning of a sense of ownership by farmers	100.0	10.0	60.0	30.0	0.0	0.0
Unification of diverse groups in the area	100.0	13.3	66.7	20.0	0.0	0.0
Freedom to raise resources	100.0	6.7	93.3	0.0	0.0	0.0
Resolution of disputes and compounding of offenses	100.0	6.7	83.3	0.0	10.0	0.0
Active involvement of all classes	100.0	6.7	60.0	13.3	20.0	0.0
D. Others						
More Farm Employment	100.0	43.3	56.7	0.0	0.0	0.0
Year-round availability of water for irrigation	100.0	20.0	80.0	0.0	0.0	0.0
Diversification of cropping pattern	100.0	36.7	63.3	0.0	0.0	0.0
Choice in deciding irrigation timings	100.0	10.0	90.0	0.0	0.0	0.0
Choice in deciding quantum of water	100.0	13.3	76.7	10.0	0.0	0.0

Diversified economic activities: Dairying	100.0	20.0	46.7	26.7	6.7	0.0
Diversified economic activities : other Non-agriculture	100.0	10.0	46.7	20.0	23.3	0.0
Transfer of power to the WUA	100.0	6.7	66.7	6.7	20.0	0.0
Transfer of power to WUA to decide water charges	100.0	6.7	73.3	0.0	20.0	0.0
Others	0.0	0.0	0.0	0.0	0.0	0.0

Source: Field survey data.

parameters, which are mentioned in Table 6.5. The varied responses of the farmers are clearly evident from the summary results presented in Table 6.5. It appears that the farmers (all groups), wage earners, livestock owners, head and tail reach farmers of the selected area have gained substantially or positively due to the formation of WUAs. Over 80-95 per cent of the farmers have expressed that these groups have gained either substantially or positively due to WUAs. On the other hand, over 40-70 per cent of the farmers have expressed that WUAs have made no impact or negative impact on tribals, lower caste/scheduled caste, youth, women group and poor people. This suggests that even after establishing WUAs, most of the benefits are going in favour of farmers, leaving very less for other groups in the selected areas.

The view of the farmers selected from the canal irrigated areas also shows that the creation of WUAs has not made any impact for the lower caste groups, landless, women and poor people (see, Table 6.6). Surprisingly, according to the perception of the farmers selected from the Lift Irrigation Users' group, the formation of Water Users Association has made substantially negative impact on tribals, lower caste and scheduled caste groups (see, Table 6.7). Since substantial contribution of capital is needed to be a member of the Lift Irrigation Users' group, the low caste and low income group would not have got the opportunity to become part of the Association. This would have substantially affected this group of farmers after the creation of WUAs. The views of the farmers selected from Pani Panchayat areas also suggest that the low income and low caste farmers have not been able to benefit from the creation of WUAs (see, Table 6.8). This is a surprising result because the objective of the Pani Panchayat was to help all the categories of the farmers, irrespective of their caste.

On the whole, the analysis on the impact of WUAs on the general economy of the area shows very mixed results. While the establishment of WUAs appeared to have made some improvement in the efficiency and equity of water use, a considerable percentage of farmers opined that the volumetric pricing has not made any impact on the efficiency and equity of water use. Though the formation of WUAs have made significant positive impact on the farmers, wage labourers, livestock owners, head reach and tail end farmers, it appears that WUAs could not make any significant impact on the group of poor, low caste (SC/ST), women and youth.

Table 6.5: Contributions of WUA to Economic Development of the Area – Results of All WUAs

Particulars	*Percentage of Sample Households Reporting*	*Substantially Positive*	*Positive*	*No Impact*	*Negative*	*Substantially Negative*
		Percentage of Reporting Households				
Village as a whole	100	15.3	54.7	30.0	0.0	0.0
Any particular religious group	100	1.3	20.7	54.7	23.3	0.0
Any particular caste	100	0.0	1.3	53.3	26.7	18.7
Any other group	100	0.0	4.7	52.0	25.3	18.0
Women	100	0.7	27.3	47.3	22.0	2.7
Poor	100	0.0	31.3	49.3	16.7	2.7
Middle income	100	0.0	46.0	41.3	12.7	0.0
Upper income	100	2.0	24.7	40.7	13.3	19.3
Large/medium farmers	100	10.7	84.7	3.3	0.7	0.7
Small/marginal farmers	100	11.3	85.3	3.3	0.0	0.0
Landless	100	0.7	37.3	48.0	14.0	0.0
Labour/wage earners	100	29.3	58.7	12.0	0.0	0.0
Livestock owners	100	32.0	55.3	8.7	4.0	0.0

Tribals	100	0.0	0.0	52.7	26.7	20.7
Upper caste	100	0.0	17.3	47.3	16.0	19.3
Lower caste	100	0.0	14.0	50.0	16.0	20.0
Scheduled castes	100	0.0	12.0	52.0	16.0	20.0
Head reach farmers	70	32.4	54.3	13.3	0.0	0.0
Tail reach farmers	70	1.9	91.4	6.7	0.0	0.0
Youth	100	0.0	5.3	64.0	18.0	12.7
Any other occupational group	100	0.0	32.7	50.7	16.7	0.0
Others	22.7	0.0	0.0	76.5	23.5	0.0

Source: Field survey data.

Table 6.6: Contributions of WUA to Economic Development of the Area - Canal Irrigation System

Particulars	*Percentage of Sample Households Reporting*	*Substantially Positive*	*Positive*	*No Impact*	*Negative*	*Substantially Negative*
		Percentage of Reporting Households				
Village as a whole	100.0	20.0	45.3	34.7	0.0	0.0
Any particular religious group	100.0	0.0	1.3	68.0	30.7	0.0
Any particular caste	100.0	0.0	2.7	69.3	28.0	0.0
Any other group	100.0	0.0	1.3	69.3	29.3	0.0
Women	100.0	0.0	4.0	66.7	29.3	0.0
Poor	100.0	0.0	10.7	60.0	29.3	0.0
Middle income	100.0	0.0	10.7	64.0	25.3	0.0
Upper income	100.0	0.0	13.3	61.3	25.3	0.0
Large/medium farmers	100.0	6.7	86.7	5.3	1.3	0.0
Small/marginal farmers	100.0	6.7	86.7	6.7	0.0	0.0
Landless	100.0	1.3	17.3	53.3	28.0	0.0
Labour/wage earners	100.0	42.7	36.0	21.3	0.0	0.0
Livestock owners	100.0	38.7	53.3	8.0	0.0	0.0

Tribals	100.0	0.0	0.0	68.0	32.0	0.0
Upper caste	100.0	0.0	9.3	62.7	28.0	0.0
Lower caste	100.0	0.0	8.0	64.0	28.0	0.0
Scheduled castes	100.0	0.0	4.0	68.0	28.0	0.0
Head reach farmers	100.0	41.3	40.0	18.7	0.0	0.0
Tail reach farmers	100.0	1.3	89.3	9.3	0.0	0.0
Youth	100.0	0.0	0.0	69.3	30.7	0.0
Any other occupational group	100.0	0.0	4.0	65.3	30.7	0.0
Others	34.7	0.0	0.0	69.2	30.8	0.0

Source: Field survey data.

Table 6.7: Contributions of WUA to Economic Development of the Area - Lift Irrigation System

Particulars	*Percentage of Sample Households Reporting*	*Substantially Positive*	*Positive*	*No Impact*	*Negative*	*Substantially Negative*
		Percentage of Reporting Households				
Village as a whole	100.0	17.8	82.2	0.0	0.0	0.0
Any particular religious group	100.0	4.4	62.2	13.3	20.0	0.0
Any particular caste	100.0	0.0	0.0	8.9	28.9	62.2
Any other group	100.0	0.0	0.0	11.1	28.9	60.0
Women	100.0	2.2	53.3	11.1	24.4	8.9
Poor	100.0	0.0	57.8	26.7	6.7	8.9
Middle income	100.0	0.0	100.0	0.0	0.0	0.0
Upper income	100.0	0.0	33.3	0.0	2.2	64.4
Large/medium farmers	100.0	0.0	97.8	0.0	0.0	2.2
Small/marginal farmers	100.0	2.2	97.8	0.0	0.0	0.0
Landless	100.0	0.0	64.4	35.6	0.0	0.0
Labour/wage earners	100.0	8.9	91.1	0.0	0.0	0.0

Livestock owners	100.0	35.6	60.0	0.0	4.4	0.0
Tribals	100.0	0.0	0.0	2.2	28.9	68.9
Upper caste	100.0	0.0	35.6	0.0	0.0	64.4
Lower caste	100.0	0.0	33.3	0.0	0.0	66.7
Scheduled castes	100.0	0.0	33.3	0.0	0.0	66.7
Head reach farmers	0.0	0.0	0.0	0.0	0.0	0.0
Tail reach farmers	0.0	0.0	0.0	0.0	0.0	0.0
Youth	100.0	0.0	0.0	48.9	8.9	42.2
Any other occupational group	100.0	0.0	100.0	0.0	0.0	0.0
Others	0.0	0.0	0.0	0.0	0.0	0.0

Source: Field survey data.

Table 6.8: Contributions of WUA to Economic Development of the Area - Pani Panchayat

Particulars	*Percentage of Sample Households Reporting*	*Substantially Positive*	*Positive*	*No Impact*	*Negative*	*Substantially Negative*
		Percentage of Reporting Households				
Village as a whole	100.0	0.0	36.7	63.3	0.0	0.0
Any particular religious group	100.0	0.0	6.7	83.3	10.0	0.0
Any particular caste	100.0	0.0	0.0	80.0	20.0	0.0
Any other group	100.0	0.0	20.0	70.0	10.0	0.0
Women	100.0	0.0	46.7	53.3	0.0	0.0
Poor	100.0	0.0	43.3	56.7	0.0	0.0
Middle income	100.0	0.0	53.3	46.7	0.0	0.0
Upper income	100.0	10.0	40.0	50.0	0.0	0.0
Large/medium farmers	100.0	36.7	60.0	3.3	0.0	0.0
Small/marginal farmers	100.0	36.7	63.3	0.0	0.0	0.0
Landless	100.0	0.0	46.7	53.3	0.0	0.0

Labour/wage earners	100.0	26.7	66.7	6.7	0.0	0.0
Livestock owners	100.0	10.0	53.3	23.3	13.3	0.0
Tribals	100.0	0.0	0.0	90.0	10.0	0.0
Upper caste	100.0	0.0	10.0	80.0	10.0	0.0
Lower caste	100.0	0.0	0.0	90.0	10.0	0.0
Scheduled castes	100.0	0.0	0.0	90.0	10.0	0.0
Head reach farmers	100.0	10.0	90.0	0.0	0.0	0.0
Tail reach farmers	100.0	3.3	96.7	0.0	0.0	0.0
Youth	100.0	0.0	26.7	73.3	0.0	0.0
Any other occupational group	100.0	0.0	3.3	90.0	6.7	0.0
Others	26.7	0.0	0.0	100.0	0.0	0.0

Source: Field survey data.

7

PROBLEMS OF WUA MEMBERS

Though the concept of participation of farmers in irrigation management is not absolutely new to the farmers in India, the rules and regulations of the formal Water Users' Association must be new to them. In an informal WUA, government devised rules are not strictly enforced for managing and distributing water, whereas the rules framed by the state agency are strictly imposed in a formal Association. Water related rules and regulations are framed in many instances without taking into account the interest of the users, which often puts constraints in the overall functioning of the institutions. In order to maintain the sustainability of WUAs, it is essential to understand the problems faced by their members. In this chapter, therefore, we try to assess the problems and the constraints faced by the sample farmers on various activities pertaining to WUAs. Suggestions made by the sample farmers to improve the overall performance of WUAs are also discussed in this chapter.

Problems Faced by the Farmers

Indian farmers have been facing a number of different problems concerning irrigation water over the years. Some of the major problems are associated with management and distribution of water, while others issues are related to pricing and investment. The problems of the farmers tend to vary from season to season and region to region as well. Therefore, we have asked the farmers to indicate their problems categorised as very major, major and occasional on 30 different parameters. Table 7.1 presents the overall responses of the sample farmers about various problems that they have faced one way or the other. It is evident from the table that the sample farmers have been facing many problems, though the major one reported by the farmers are very small. High cost of maintenance, high cost of electricity, lack of government funding, inadequate field channels, lack of training to staff/members and lack of freedom to determine the water rates appeared to be some of the major problems faced by the sample farmers in the selected areas. Over 40 per cent of the farmers have expressed that high cost of maintenance and cost of electricity are the major problems that they encountered, whereas inadequate field channels, lack of financial support, lack of training to staff/members have been reported as the major problems by about 20 to 40 per cent of the farmers. Most of the problems are appeared to be associated with investment and price of irrigation water, which is always a major problem of the farmers even before the formation of WUAs.

The problems expressed by the canal water users are somewhat similar to the overall results reported earlier (see, Table 7.2). Over 50 per cent of the farmers have expressed that inadequate and high cost of maintenance, inadequate field channels, lack of government support (funding) and lack of training to staff/members are the major problems. However, unlike the overall results, high cost of electricity is not a major problem for canal water users, because they need not use electricity for irrigation purposes.

The responses of the farmers selected from the Lift Irrigation Users' group are totally different from the one that we have seen with the Canal Water Users' group. About 45 to 60 per cent of the farmers have expressed that lack of government funding and high cost of electricity are the major problems that they have been encountering after the formation of WUAs (see, Table 7.3). Other than these two, the farmers belonging to lift irrigation group have not faced any other major problems. Since the cost of electricity is one of the major costs for the users of lift irrigation, this has been reported as the major problem by the farmers. High cost of electricity and lack of financial support (start-up capital) are also expressed as the major problems by the farmers selected from Pani Panchayat group (see, Table 7.4). This is expected because farmers need to spend more money for electricity because water is lifted from wells for irrigating crops under Pani Panchayat system.

On the whole, high cost of electricity and lack of government funding have imaged as the major problems of the farmers selected from both lift irrigated areas and also in Pani Panchayat areas. High cost of maintenance, inadequate field channels, lack of government funding and lack of freedom to determine the water rates seem to be major problems faced by the Canal Water Users.

Table 7.1: Problems Faced by the Members of WUAs on Various Aspects – Results of All WUAs

Problems	*Percentage of Sample Households Reporting*	*Very Major*	*Major*	*Light / Occasional*	*None / No*
		Percentage of Reporting Households			
1. Inadequate maintenance	100.0	2.0	33.3	28.0	36.7
2. High cost of maintenance	100.0	14.7	26.0	26.0	33.3
3. Lack of members cooperation	100.0	2.0	5.3	48.7	44.0
4. Non availability of water at the canal	100.0	1.3	0.7	67.3	30.7
5. Poor quality of water	100.0	0.0	1.3	50.0	48.7
6. Conflict among members about timing of water	100.0	0.7	1.3	23.3	74.7
7. Conflict among members about quantity of water	100.0	0.7	1.3	22.0	76.0
8. Conflict among members about pricing of water	100.0	1.3	4.7	20.0	74.0
9. Inadequate field channels	100.0	4.7	24.7	20.0	50.7
10. Lack of government support/ funding	100.0	6.7	44.0	22.7	26.7

11. High cost of electricity	100.0	18.7	22.7	14.0	44.7
12. Lack of financial support (start-up)	100.0	3.3	32.0	18.0	46.7
13. High cost of tube-well repairing	70.0	0.0	0.0	7.6	92.4
14. Water table receding fast	100.0	0.0	1.3	29.3	69.3
15. Lack of mechanism to control water use	100.0	0.7	16.7	24.0	58.7
16. Lack of quality planting materials	100.0	0.0	3.3	34.0	62.7
17. Lack of consensus in deciding cropping pattern	100.0	0.7	1.3	23.3	74.7
18. Complaints from tail reach farmers	70.0	1.0	7.6	48.6	42.9
19. Lack of training to staff/ members	100.0	0.7	36.7	30.0	32.7
20. Non payment of water charges	100.0	0.0	11.3	44.0	44.7
21. Lack of investment credit with farmers	100.0	6.0	20.0	36.0	38.0
22. Problems in devising water distribution rules	100.0	2.7	8.0	38.0	51.3
23. Lack of member cooperation	100.0	0.0	1.3	44.7	54.0
24. Problems during the period when water is very scarce	100.0	1.3	13.3	54.7	30.7
25. The margins earned by the association is too meager	100.0	0.0	10.7	40.7	48.7

26. Limited control over water flow	100.0	3.3	18.0	25.3	53.3
27. Lack of leadership	100.0	1.3	0.7	18.7	79.3
28. Lack of freedom to determine water rates	100.0	1.3	19.3	23.3	56.0
29.Lack of member's willingness to take up management functions	100.0	5.3	15.3	23.3	56.0
30. Other	28.0	0.0	0.0	14.3	85.7

Source: Field survey data.

Table 7.2: Problems Faced by the Members of WUAs on Various Aspects - Canal Irrigation System

Problems	*Percentage of Sample Households Reporting*	*Very Major*	*Major*	*Light / Occasional*	*None / No*
		Percentage of Reporting Households			
1. Inadequate maintenance	100.0	4.0	57.3	10.7	28.0
2. High cost of maintenance	100.0	24.0	33.3	8.0	34.7
3. Lack of members cooperation	100.0	1.3	6.7	64.0	28.0
4. Non availability of water at the canal	100.0	2.7	1.3	73.3	22.7
5. Poor quality of water	100.0	0.0	2.7	53.3	44.0
6. Conflict among members about timing of water	100.0	1.3	2.7	24.0	72.0
7. Conflict among members about quantity of water	100.0	1.3	2.7	20.0	76.0
8. Conflict among members about pricing of water	100.0	2.7	5.3	14.7	77.3
9. Inadequate field channels	100.0	9.3	48.0	16.0	26.7
10. Lack of government support/ funding	100.0	4.0	50.7	17.3	28.0

11. High cost of electricity	100.0	0.0	6.7	4.0	89.3
12. Lack of financial support (start-up)	100.0	6.7	41.3	6.7	45.3
13. High cost of tube-well repairing	100.0	0.0	0.0	1.3	98.7
14. Water table receding fast	100.0	0.0	2.7	8.0	89.3
15. Lack of mechanism to control water use	100.0	1.3	33.3	8.0	57.3
16. Lack of quality planting materials	100.0	0.0	6.7	29.3	64.0
17. Lack of consensus in deciding cropping pattern	100.0	1.3	0.0	13.3	85.3
18. Complaints from tail reach farmers	100.0	1.3	10.7	65.3	22.7
19. Lack of training to staff/ members	100.0	1.3	53.3	10.7	34.7
20. Non payment of water charges	100.0	0.0	22.7	52.0	25.3
21. Lack of investment credit with farmers	100.0	12.0	32.0	24.0	32.0
22. Problems in devising water distribution rules	100.0	5.3	12.0	33.3	49.3
23. Lack of member cooperation	100.0	0.0	1.3	54.7	44.0
24. Problems during the period when water is very scarce	100.0	2.7	24.0	46.7	26.7

25. The margins earned by the association is too meager	100.0	0.0	20.0	29.3	50.7
26. Limited control over water flow	100.0	6.7	34.7	10.7	48.0
27. Lack of leadership	100.0	2.7	1.3	12.0	84.0
28. Lack of freedom to determine water rates	100.0	2.7	38.7	8.0	50.7
29. Lack of member's willingness to take up management Functions	100.0	10.7	28.0	13.3	48.0
30. Other	34.7	0.0	0.0	11.5	88.5

Source: Field survey data.

Table 7.3: Problems Faced by the Members of WUAs on Various Aspects - Lift Irrigation System

Problems	*Percentage of Sample Households Reporting*	*Very Major*	*Major*	*Light / Occasional*	*None / No*
		Percentage of Reporting Households			
1. Inadequate maintenance	100.0	0.0	0.0	53.3	46.7
2. High cost of maintenance	100.0	4.4	13.3	60.0	22.2
3. Lack of members cooperation	100.0	0.0	0.0	40.0	60.0
4. Non availability of water at the canal	100.0	0.0	0.0	75.6	24.4
5. Poor quality of water	100.0	0.0	0.0	55.6	44.4
6. Conflict among members about timing of water	100.0	0.0	0.0	35.6	64.4
7. Conflict among members about quantity of water	100.0	0.0	0.0	37.8	62.2
8. Conflict among members about pricing of water	100.0	0.0	0.0	35.6	64.4
9. Inadequate field channels	100.0	0.0	2.2	6.7	91.1
10. Lack of government support/ funding	100.0	15.6	31.1	37.8	15.6

11. High cost of electricity	100.0	28.9	31.1	40.0	0.0
12. Lack of financial support (start-up)	100.0	0.0	8.9	46.7	44.4
13. High cost of tube-well repairing	0.0	0.0	0.0	0.0	0.0
14. Water table receding fast	100.0	0.0	0.0	71.1	28.9
15. Lack of mechanism to control water use	100.0	0.0	0.0	60.0	40.0
16. Lack of quality planting materials	100.0	0.0	0.0	60.0	40.0
17. Lack of consensus in deciding cropping pattern	100.0	0.0	0.0	44.4	55.6
18. Complaints from tail reach farmers	0.0	0.0	0.0	0.0	0.0
19. Lack of training to staff/ members	100.0	0.0	20.0	77.8	2.2
20. Non payment of water charges	100.0	0.0	0.0	40.0	60.0
21. Lack of investment credit with farmers	100.0	0.0	0.0	66.7	33.3
22. Problems in devising water distribution rules	100.0	0.0	0.0	60.0	40.0
23. Lack of member cooperation	100.0	0.0	0.0	42.2	57.8
24. Problems during the period when water is very scarce	100.0	0.0	0.0	80.0	20.0

25. The margins earned by the association is too meager	100.0	0.0	2.2	71.1	26.7
26. Limited control over water flow	100.0	0.0	2.2	60.0	37.8
27. Lack of leadership	100.0	0.0	0.0	42.2	57.8
28. Lack of freedom to determine water rates	100.0	0.0	0.0	48.9	51.1
29. Lack of member's willingness to take up management Functions	100.0	0.0	4.4	46.7	48.9
30. Other	4.4	0.0	0.0	100.0	0.0

Source: Field survey data.

Table 7.4: Problems Faced by the Members of WUAs on Various Aspects - Pani Panchayat

Problems	*Percentage of Sample Households Reporting*	*Very Major*	*Major*	*Light / Occasional*	*None / No*
		Percentage of Reporting Households			
1. Inadequate maintenance	100.0	0.0	23.3	33.3	43.3
2. High cost of maintenance	100.0	6.7	26.7	20.0	46.7
3. Lack of members cooperation	100.0	6.7	10.0	23.3	60.0
4. Non availability of water at the canal	100.0	0.0	0.0	40.0	60.0
5. Poor quality of water	100.0	0.0	0.0	33.3	66.7
6. Conflict among members about timing of water	100.0	0.0	0.0	3.3	96.7
7. Conflict among members about quantity of water	100.0	0.0	0.0	3.3	96.7
8. Conflict among members about pricing of water	100.0	0.0	10.0	10.0	80.0
9. Inadequate field channels	100.0	0.0	0.0	50.0	50.0
10. Lack of government support/ funding	100.0	0.0	46.7	13.3	40.0
11. High cost of electricity	100.0	50.0	50.0	0.0	0.0

12. Lack of financial support (start-up)	100.0	0.0	43.3	3.3	53.3
13. High cost of tube-well repairing	100.0	0.0	0.0	23.3	76.7
14. Water table receding fast	100.0	0.0	0.0	20.0	80.0
15. Lack of mechanism to control water use	100.0	0.0	0.0	10.0	90.0
16. Lack of quality planting materials	100.0	0.0	0.0	6.7	93.3
17. Lack of consensus in deciding cropping pattern	100.0	0.0	6.7	16.7	76.7
18. Complaints from tail reach farmers	100.0	0.0	0.0	6.7	93.3
19. Lack of training to staff/ members	100.0	0.0	20.0	6.7	73.3
20. Non payment of water charges	100.0	0.0	0.0	30.0	70.0
21. Lack of investment credit with farmers	100.0	0.0	20.0	20.0	60.0
22. Problems in devising water distribution rules	100.0	0.0	16.7	10.0	73.3
23. Lack of member cooperation	100.0	0.0	3.3	23.3	73.3

24. Problems during the period when water is very scarce	100.0	0.0	6.7	36.7	56.7
25. The margins earned by the association is too meager	100.0	0.0	0.0	23.3	76.7
26. Limited control over water flow	100.0	0.0	0.0	10.0	90.0
27. Lack of leadership	100.0	0.0	0.0	0.0	100.0
28. Lack of freedom to determine water rates	100.0	0.0	0.0	23.3	76.7
29. Lack of member's willingness to take up management functions	100.0	0.0	0.0	13.3	86.7
30. Other	46.7	0.0	0.0	7.1	92.9

Source: Field survey data.

Table 7.5: Assessment about the Success of WUAs and its Financial Health

Particulars	*Percentage of Sample Households Reporting*	*Very Successful*	*Successful*	*Satisfactory*	*Poor*
		Percentage of Reporting Households			
Assessment about the Success of WUA					
Canal Irrigation System	98.7	0.0	28.4	66.2	5.4
Lift Irrigation System	100.0	6.7	46.7	46.7	0.0
Pani Panchayat	100.0	13.3	36.7	13.3	36.7
Overall	99.3	4.7	35.6	49.7	10.1
Assessment about the Financial Health of WUA					
Canal Irrigation System	96.0	0.0	2.8	93.1	4.2
Lift Irrigation System	100.0	2.2	20.0	77.8	0.0
Pani Panchayat	98.0	0.7	12.9	74.1	12.2
Overall	98.0	0.7	12.9	74.1	12.2

Source: Field survey data.

Assessment on the Success of WUAs

After having assessed the major problems faced by the members of the WUAs, we have tried to assess the success of WUAs, using the responses provided by the sample farmers. In order to assess the success of WUAs, farmers were asked to indicate their perceptions about two points: (a) overall assessment about the success of WUAs and (b) assessment about the financial health of WUAs. Four options viz., very successful, successful, satisfactory and poor were given to the sample farmers to assess the success of WUAs.

The summary of the farmers' responses about the assessment of WUAs are presented in Table 7.5. Except 10 per cent of the farmers, all other farmers have expressed that overall performance of WUAs is either successful or satisfactory. Considerable number of farmers belonging to tail-end WUAs and poor caste groups have expressed that they could not see any big change in water supply and management even after the formation of WUAs. The assessment of the farmers about the success of WUAs varies with the farmers selected from different types of WUAs. While about 36 per cent of the farmers selected from Pani Panchayat areas have expressed that the overall success is poor, the same is not true with the assessment of the farmers selected from canal and lift water users groups. Since the overall performance of the Pani Panchayat has deteriorated over the years due to various reasons, the farmers are not willing to admit that their WUA is successful. Unlike the responses of the farmers selected from Pani Panchayat area, over 95 per cent of the farmers selected from canal and lift irrigation water users groups expressed that the overall success of WUAs is either satisfactory or successful.

The financial health of WUAs appears to be in good shape from the responses of the farmers. Over 90 per cent of the farmers from canal and lift irrigation water users' group have expressed that the financial health of their WUAs is either successful or satisfactory. As expected, about 12 per cent of the farmers selected from Pani Panchayat areas have expressed that the financial condition of their institution is poor. Since many of the members opted out from the Pani Panchayat institutions, the financial health of the institutions must have weakened over the years.

Farmers' Suggestions to Improve the Functioning of WUAs

It is seen from the above that the users have encountered a number of problems at different levels in carrying out the works relating to WUAs.

Table 7.6: Farmers' Suggestions to Improve the Functioning of WUA

Canal Irrigation System	*Lift Irrigation System*	*Pani Panchayat*
1. Proper maintenance work should be carried out in time.	1. Electricity tariff should be reduced.	1. Electricity tariff should be reduced.
2. Maintenance cost should be minimized.	2. Government should support the association.	2. Government should support the association
3. Government should support the system by adequate funding.	3. Maintenance cost should be lower down.	3. Financial support (start-up capita) should be provided by government.
4. Field channels should be expanded and old field channels should be repaired.	4. Proper training should be given to the staff members.	4. Proper maintenance work should be carried out.
5. Proper training should be given to the staff and members.	5. Institutional credit should be made available to farmers at cheaper rate and in time.	
6. Institutional credit should be made available to farmers at cheaper rate and in time.	6. Water scarcity should be solved by proper planning	
7. The release of water should be as per farmers' demand.		
8. Water rates should be fixed by the farmers.		
9. A proper mechanism should be developed to control water use		
10. Water scarcity should be solved by proper planning		

Source: Field survey data.

Though the problems encountered by the sample farmers selected from three types of WUAs are not the same, problems relating to investment, costs and pricing as well as management of water are commonly seen in all the selected areas. In order to improve the overall functioning of WUAs, we have asked the sample farmers to give important suggestions. In this section, we present only the major suggestions that come out from the sample farmers.

It is evident from Table 7.6 that the suggestions made by the sample farmers are not the same. Under canal irrigation system, the sample farmers have suggested 10 major points to improve the overall functioning of WUAs. Majority of the suggestions are one way or the other related to maintenance work of the channels. The sample farmers feel that the maintenance work has to be carried out in time and government should continue to support the maintenance work. While suggesting training for staff and members of WUAs, farmers have also pointed out that the timings of release of water and fixation of water rates should be with the users' group.

The major suggestions of the farmers selected from the areas of lift irrigation and Pani Panchayat are somewhat different from the suggestions made by the farmers selected from the canal irrigated areas. In both the lift irrigated areas as well as in Pani Panchayat areas, farmers have suggested that the electricity cost should be reduced and government should also support the institutions by providing initial start-up capital. As reported earlier, since electricity cost is the major item of the expenditures, majority of the farmers feel that the functioning of these Associations can be improved, if the cost of electricity is reduced substantially.

On the whole, though the sample farmers have expressed that the overall achievement of WUAs are either satisfactory or successful, the sample farmers have been facing a number of different problems in improving the overall performance of WUAs. The problems faced by the canal water users are somewhat different from the users selected from Pani Panchayat and lift irrigated areas.

8

SUBSTANCE AND SUGGESTIONS

Irrigation water plays a paramount role in increasing the growth of agricultural output. It increases the cropping intensity, changes the cropping pattern in favour of high value crops, encourages the adoption of technological inputs as well as machineries, all of which one way or the other help to augment the crop output. Besides directly benefiting the farmers, it also indirectly benefits the non-farming community substantially. While benefiting the landless agricultural labourers in terms of increased employment opportunities and wage rate, irrigation helps to reduce the rural poverty in a sustained manner. In view of the importance of irrigation development in agriculture and other sectors, the policy makers in India have been giving increased thrust for the development of irrigation since independence. Despite significant achievements in the development of irrigated area, the overall performance of irrigation sector has not been very appreciable in India. While the water use efficiency is very low (ranging from 30-40 per cent under surface method of irrigation) in India as compared to other countries, the financial performance of the State managed irrigation sector has been deteriorating over the years. The percent of recovery rate (ratio of irrigation revenue to operation and maintenance cost) of irrigation and multipurpose river valley projects, which are completely managed by the State agency, has declined from 36.4 per cent in the 1976-77 to as low as 5.7 per cent in 1999-2000.

Though there are many reasons for the dreadful performance of the irrigation sector, significant increase in operation and maintenance (O&M) expenditures of the irrigation sector and unrevised irrigation water rates charged from farmers are often cited as the main reasons for the poor financial performance of the publicly funded irrigation sector. It has been argued recently that the poor water use efficiency, which exists across different regions in India, is the result of low water price fixed for different crops. Poor participation of farmers in irrigation management related activities is also cited as one of the major reasons for the deteriorating performance of irrigation sector in India. Researchers suggest that the poor performance of irrigation sector can be improved substantially by extensively involving the users (farmers) in the management activities of the irrigation sector.

Involving farmers in water management activities is not a new idea as far as India is concerned. In the name of kudimaramath in South India, the users (farmers) were involved in tank repairs, management works, distilling of tanks, removal of weeds, distribution of water and collecting revenue even during the 17th and 18th century, much before the British rule. Unfortunately, due to various reasons, the users' participation in water management has declined drastically over the years. Now, partly because of pressure from donor agencies such as the World Bank and others, the users' participation in irrigation water has become a widespread strategy in all countries in Asia including India at least from the mid-eighties. The ultimate aim of the decentralized water management with predominant users' participation is to improve the overall performance of irrigation sector. In India, much emphasis was given for users' participation only after the announcement of the first National Water Policy (1987), where gradual involvement of farmers in system management was advocated. The second National Water Policy (2002) also reiterates the importance of farmers' participation in irrigation management activities.

What is the ideal condition under which the local level collective action would be successful in managing the common property resources? This is one of the issues of the common property resource management, which has attracted a large number of researchers around the world in recent years. Many studies are available focusing on this issue, but the results of the studies are not the same. While one group of researchers argue that the cooperative behaviour is difficult to sustain in an environment which has large heterogeneous agents, the other group of researchers argue that heterogeneity does not necessarily matter in bringing collective participation in the commons. Specific studies have also been carried out relating the performance of community based irrigation systems and group heterogeneity using data from different countries. However, not many studies are available focusing on this issue in the Indian context. In the present study, besides looking at the heterogeneity of the members of WUAs, we also make an attempt to see as to how the heterogeneity in physical characteristics such as location of these Associations and source of water used by them affect the participation of farmers in water management activities and other impact parameters. This study has been carried out in Maharashtra State, which has made rapid progress in establishing WUA since mid-nineties.

Objectives of the Study

1. To study what extent WUAs have contributed towards regular water supply, efficiency in water use, collection of water charges, and operation and maintenance of the water delivery systems.
2. To study the impact of WUAs on cropping pattern, input use and improvement in productivity of crops as well as in the socio-economic well being of the beneficiaries.
3. To study what extent the devolution of power to WUAs has taken place in the State.
4. To identify the constraints in the effective implementation of PIM and ways to remove the same.
5. To suggest policy recommendations for the effective implementation of PIM.

Scope and Information Analysis

The study has been carried out in Maharashtra State, which has made policy decision to transfer the whole irrigation system management to users (farmers) group since the mid-nineties. In order to provide legal recognition to the Water Users Association, the state has also enacted an Act which is known as Maharashtra Management of Irrigation System by Farmers Act, 2005. As a result of these initiatives, there has been a significant progress in the number of WUAs functioning/registered/ proposed since September 1996. While the number of fully functioning WUAs increased from 100 in September 1996 to 774 in July 2005, the area controlled by these increased from about 43.68 thousand hectares to 2.51 lakh hectares during this period. That is, the area covered by them has increased nearly by six times during this period. The total number of WUAs (signed/registered/proposed) increased from 457 in September 1996 to 4051 as of July 2005 and the area covered by them increased from about 1.69 lakh hectares to about 15.41 lakh hectares in July 2005.

Since the performance of WUAs functioning at different locations using different sources of water is expected to be different because of heterogeneity in physical and other characteristics, altogether 10 WUAs have been selected using the available data as well as discussions with the officials of the irrigation department in Maharashtra State. Of the 10

WUAs, five WUAs have been selected from canal command areas, three from Lift Irrigation System (LIS) and two from Pani Panchayat scheme. Our aim is to study the functioning of WUAs at different locations of the canal command areas and therefore, one WUA from head reach of the canal and two each from middle and tail end of the canal command areas have been selected. Similarly, three WUAs established in lift irrigated areas using water from different sources namely river, well and backwater have been selected. However, the same method has not been followed while selecting WUAs from Pani Panchayat areas as it mostly uses well water.

Data has been collected from both the sample farmers as well as the records of all the 10 WUAs selected for the study. A total 150 sample farmers - 75 farmers from canal irrigation system, 45 farmers from lift irrigation system and 30 from Pani Panchayat areas - have been selected randomly for the study. Out of 75 farmers selected from canal irrigation system, 15 farmers have been selected from WUA functioning in the head reach of the canal command, 30 from the middle region and another 30 from the tail end of the canal command area. From lift irrigation system, 15 farmers have been selected from each WUA operating under river water, well water and backwater. A total 30 sample farmers have been selected from the two Pani Panchayat WUAs. The sample farmers have been selected randomly using the list of farmers provided by each WUA, covering different size groups namely small, medium and large.

The major objective of the study is to find out the overall impact of WUAs selected from different locations on various parameters. Therefore, field level data has been collected not only from the sample farmers but also from each selected WUA. Data on water price, financial recovery, cropping pattern and other institutional related information have been collected from the office records of each WUA to study the impact of it on various parameters. Data on different aspects pertaining to the study has been collected from the sample farmers in order to find out the ground level impact of WUAs on different parameters.

Revelations

Profile of WUAs

The analysis on the profile of ten water users' association selected from different locations shows varied results. The overall working conditions of WUAs established in the canal irrigated areas and lift irrigated areas appear to be better. But the same is not fully true with

WUAs established in Pani Panchayat areas. Considerable increase in gross cropped area and irrigated area has been reported due to the formation of WUA. The financial status of the two WUAs – one from lift irrigation WUA and another one from Pani Panchayat – are not in good shape because of increased operation and maintenance expenditures.

Characteristics of the Sample Households

The personal characteristics of the sample households selected from different WUAs are not the same. The average age of the respondents selected from canal irrigation system is relatively lower (43.16) as compared to the farmers selected lift irrigation system (55.98) and Pani Panchayat system (53.03). Similar kind of differences is also observed in the farming experience of the sample farmers. Though there are differences in the family size of sample households selected by different WUAs, it is positively related with farm size in all WUAs.

Differences are also noticed in the position that the respondents occupy in their respective WUAs across all WUAs. Considerable percentage of farmers belonging to large farm size category are occupying some respectable position (either Chairman or Vice-Chairman or Managing Committee Member) in WUAs working under canal and lift irrigation system. This is not true in the case of Pani Panchayat, where almost all the sample farmers are just members of the Water Users' Association.

The average land holding size of the household is almost same (varies from 2.43 ha to 2.75 ha) across all WUA's. While the average holding size comes to 2.43 ha for the farmers selected from canal WUAs, the same comes to 2.75 ha and 2.62 ha respectively for the farmers selected from lift irrigation societies and pani panchayat system.

The level of irrigation is not the same across different farm size categories and also between the households selected from different WUAs. The share of irrigated area to cropped area of the farmers selected from the canal and lift irrigation water users' group is 83 and 100 percent respectively, whereas the same is only 45 percent for those farmers selected from Pani Panchayat system. Since water scarcity is always higher in Pani Panchayat areas, the percentage of irrigated area is very less among the farmers.

The level of irrigation across different farm size groups is also not the same, except the farmers selected from lift irrigation societies. While the level of irrigation comes to 100 percent for all the three size categories of farmers selected from lift irrigation society, this is not true in the case of farmers selected from canal and Pani Panchayat areas. The coverage of irrigation comes to 93.52 percent for small farmers, but the same is 79.89 per cent for the large size farmers selected from the canal irrigated areas. Similarly, the coverage of irrigation comes to 54.86 per cent for the small size farmers, whereas the same comes to only 32.92 per cent for the large farm size in Pani Panchayat areas.

Role and Functions of WUAs

The role of various functionaries and socio-economic groups in the functioning of the Water Users' Associations is studied using the responses of the sample farmers. The overall results show that the very active participation is very low from different stakeholders in various activities pertaining to WUAs. Only 16 per cent of the respondents have expressed that the Chairman's role has been very active. Around 30 per cent of the respondents expressed that the head reach farmers have been involved very actively in various activities of WUAs. While the role of Panchayat institutions is found to be almost none, the role of farmers group (all size of farmers) and agricultural labourers is found to be relatively better than others.

Canal Irrigation

According to the responses of the farmers selected from canal irrigated areas, no groups/functionaries appeared to have played active role in the overall functioning of WUAs. Considerable percentage of farmers have expressed that the role of the office bearers of WUAs is active. Large size farmers, government officials, chairman and secretary of WUAs as well as tail end farmers reported to have played active role in the functioning of WUAs working in the canal command areas. The village institutions such as Panchayat, cooperative credit society, caste groups, etc., reported to have played very limited role in the overall functioning of WUAs.

Lift Irrigation

The role played by different stakeholders in improving the performance of WUAs established in lift irrigated areas seems to be better. Either "very active" or "active" role played by the members of WUAs is found to be much better than that of the responses of the farmers selected from canal command areas. The role of the village institutions in the overall functioning of the lift irrigation societies is completely absent.

Pani Panchayat

Pani Panchayat is formed with the support of an NGO in order to protect the livelihood of the farmers living in the rainfed areas. With direct control of the government the Panchayat's role is very negligible in regulating the performance of WUAs. Farmers' responses show that the "very active" role played by the different stakeholders is not much different from other two WUAs. Surprisingly, even the members' role is reported to be "very poor" in the overall functioning of WUAs. Since the members have lost interest on Pani Panchayat system because of inadequate water supply, restrictions on cultivating water-intensive crops and other reasons, the "active role" played by the members declined over the years.

Devolution of Power

The farmers' responses about the changes in the devolution of powers before and after the establishment of WUAs overall show very encouraging results. The sample farmers could realise considerable changes in the devolution of power from the state agency to the users group after establishing WUAs. The control of state agency was reported to be over 60 to 100 per cent in different subject areas before the establishment of WUAs. This has now totally changed in most cases. Around 50 per cent of the sample farmers feel that the State control still exists particularly in the case of planning, releasing and distribution of water.

Canal Irrigation

Significant changes in the devolution of power have been noted by the farmers of the canal command areas. Before the formation of WUAs in the canal command areas, the state agency (irrigation department) was controlling almost all the aspects pertaining to irrigation water. This has changed dramatically after the establishment of WUAs, except in the case of assessment and release of water. About 99 per cent of farmers expressed that the State control was very significant in different aspects of water before the formation of WUAs, but now only 26 per cent of the farmers expressed that the state control still exists. Majority of the farmers have noticed changes in the devolution of power in cases like pricing of water, collection of dues and maintenance.

Lift Irrigation

Since lift irrigation societies are formed through farmers own initiatives and resources under the Act of lift irrigation cooperative societies, over 94 per cent of the farmers selected from the lift irrigation water users group expressed that they enjoy full autonomy in every aspects of irrigation water, except in the case of pricing of water.

Pani Panchayat

Pani Panchayat is established with the support of a non-governmental organisation so as to protect the livelihood of the farmers under certain unusual conditions. The devolution of power has been part of Pani Panchayat since the inception. Except in the case of pricing and distribution of water, the control of state agency is very negligible in all other cases. Since electricity is supplied by the state agency and it is essential for supply of water in time for crop cultivation, over 45 to 66 per cent of farmers belonging to Pani Panchayat have expressed that they could not see any change in the devolution of power, after its formation.

Impact of WUAs on Cropping Pattern and Irrigated Area

The impact of Water Users' Association on farm economy has also been studied. Overall positive changes have been noted in cropping pattern as well as in area under irrigation after the formation of WUAs. While the area allotted for kharif season has marginally declined, the same has increased marginally in rabi and summer seasons after the formation of WUAs. Since foodgrain crops do not give much profit, farmers have reduced area under foodgrain crops and have increased the area under non-foodgrain crops such as oilseeds and vegetables. After the formation of WUAs, the area under foodgrain crops has declined by almost 35 per cent in kharif season and around six per cent in rabi season. Interestingly, area allotted for summer crops has increased by about four times (from 0.11 ha to 0.44/ha/household) after the establishment of WUAs.

Overall area under irrigated crop has increased after the formation of WUAs. The share of irrigated crop area was about 73 per cent before the formation of WUAs, but the same increased to nearly 92 after establishing WUAs. Significant increase in the percentage of irrigated area to cropped area has been noted in summer season as compared to other two seasons. The increase in irrigated area was about 31 per cent in summer season, whereas the same was only about 24 and 17 per cent respectively in kharif and rabi seasons.

Canal Irrigation

Positive changes in cropping pattern and irrigated area have also been observed among the farmers selected from canal irrigated areas. The area allotted for better value crops such as vegetables, etc., has increased in all the three seasons, including summer season. The share of irrigated cropped area has increased from 77 per cent to over 94 per cent after the formation of WUAs. This increase was possible because of better water management by farmers in canal command areas after the creation of WUAs.

Lift Irrigation

Positive changes in cropping pattern and irrigated area are also seen among the farmers selected from lift irrigated areas. Farmers have allotted more area for better value crops than foodgrain crops after the formation of societies. The area cultivated during summer season has increased from 0.09 ha/household to 0.86 ha/household.

Pani Panchayat

The performance of Pani Panchayat has been deteriorating over the years, both cropping pattern and area under irrigation has changed significantly among the farmers selected from Pani Panchayat areas. The coverage of irrigation was only 1.21 per cent of the cropped area before the creation of Pani Panchayat, but now this has increased to over 71 per cent.

Impact of WUAs on Input Use

The overall responses of the farmers show significant positive changes in the adoption of various inputs after the creation of WUAs. Over 24 to 37 per cent of farmers have expressed that they could increase HYV/improved seed and fertilisers to a "large extent" after the establishment of WUAs. Since local seed does not provide higher yield, over 96 per cent of farmers could reduce the same after the establishment of WUAs. Significant reduction in the use of bullock labour has also been reported by the farmers, because of added use of machineries such as tractors, etc. Over 90 per cent of farmers expressed that the cost of cultivation (irrigation and other costs) has increased because of change in input use after the creation of WUAs.

Farmers selected from canal water users' group have also reported that they could increase the use of various inputs after the establishment of WUAs. Majority of the farmers reported to have increased the adoption of HYV seed, fertilisers, pesticides, etc., in crop cultivation. Because of change in cultivation due to better availability of water, farmers also reported to have significantly reduced local seed, bullock labour, etc. The same kind of trend is also seen in the use of inputs among the farmers selected from lift irrigation water users' group as well as from Pani Panchayat group.

Impact of WUAs on Productivity of Crops

On an average, over 90 per cent of farmers have expressed that they could realise either "large increase" or "increase" in productivity of major crops such as bajra, wheat, jowar, groundnut, sugarcane, cauliflower and other vegetable crops, etc, after the establishment of WUAs. Though the response of the farmers about productivity increase varies across crops, the overall trend is that the farmers are able to increase the productivity of major crops after the establishment of WUAs. This is expected because the establishment of WUAs has allowed the farmers to use higher quantity of yield increasing technological inputs, which must have helped the farmers to get higher productivity from different crops.

Considerable increase in productivity of major crops has been reported by majority of the farmers selected from different WUAs. About 30 percentage of farmers selected from the Canal Water users' group reported that the productivity of crops (especially in groundnut, sugarcane and onion) seems to have declined after establishment of WUAs. This reduction in productivity is not mainly because of supply of water but due to diseases and other problems. However, among the three types of WUAs, none of the farmers selected from lift irrigation users' group and Pani Panchayat reported any decrease in productivity of crops.

Impact of WUAs on General Economy

The impact of WUAs on the efficiency, equity and social cohesion has also been studied using the responses of the farmers. The overall results appear to suggest that farmers' responses on different parameters are not the same. On the issue of efficiency, though over 90 per cent of the farmers have indicated that they could see "highly positive" or "positive" changes in adequate water availability and increase in irrigated area, about 30 per cent of the farmers have expressed that they have realised "no impact" on the timely availability of water, after the creation of WUAs. This is particularly true in the case of farmers who are part of the water users' association located at the tail end of canal command area.

With regard to the issue of equity, about 30 per cent of farmers expressed that the volumetric pricing and voting rights for all WUA members have made “no impact” or made “negative impact” on the overall development of the economy in the selected area. The survey results suggest that WUAs have made considerable impact on sense of ownership, freedom to raise resources and unification of diverse groups in the area. However, it appears from the responses of the farmers that WUAs have not made significant impact on solving disputes and bringing active involvement of all classes of farmers.

The creation of WUAs appears to have made considerable impact on the farm employment, diversification of cropping pattern and other economic activities (dairying, etc.) as well as transfer of power to WUAs. However, nearly 30 per cent of the farmers expressed that they could see “no impact” on deciding irrigation timings as well as quantum of water supply after the creation of WUAs.

Canal Irrigation

The overall responses of the farmers on the efficiency, equity and social cohesion are almost matching with the responses of the farmers selected from the canal water users’ group. Nearly 50 per cent of the farmers belonging to the canal water users’ group have felt that WUAs have made “no impact” or “negative impact” especially on parameters such as (a) timely availability of water, (b) maintenance of irrigation structure, (c) reduction in cost of maintenance, (d) pricing of irrigation water, (e) unification of diverse groups in the area, (f) resolution of disputes and compounding of offences, (g) round the year availability of irrigation, (h) choice of irrigation times, and (i) choice in deciding quantum of water.

Lift Irrigration

The responses of the farmers selected from the lift irrigation users’ group show a much better picture about the overall impact of WUAs on the general economy. Over 90 per cent of the farmers have expressed that they could realise either “highly positive” or “positive” impact on all the parameters after the formation of WUAs, except parameters such as (a) resolution of disputes and compounding of offences, and (b) diversified economic activities. Since the farmers from lift irrigation group have invested relatively more than the other group of farmers, it must have motivated the farmers to get involved intensively in all activities related to WUAs.

The responses of the sample farmers selected from Pani Panchayat group also show relatively better picture about the impact of WUAs on the general economy of the area, barring a few parameters.

Contribution of WUAs to Economic Development

The contributions of WUAs to the economic development of the selected area have also been studied. Farmers (all groups), wage earners, livestock owners, head and tail end farmers of the selected area appeared to have gained substantially or positively due to the formation of WUAs. Over 80-95 per cent of the farmers have expressed that these groups have gained either substantially or positively due to WUAs. On the other hand, over 40-70 per cent of the farmers have expressed that WUAs have made "no impact" or "negative impact" on tribals, lower caste/scheduled caste, youth, women group and poor people.

The view of the farmers selected from the canal irrigated areas also shows that the creation of WUAs has not made any impact on the lower caste groups, landless, women and poor people. According to the view of the farmers selected from the Lift Irrigation Users' group, the formation of Water Users Association has made "substantially negative" impact on tribals, lower caste and scheduled caste groups. Since substantial contribution of capital is needed to be a member of the lift irrigation users' group, the low caste and low income group farmers would not have got the opportunity to become part of these Association. The views of the farmers selected from Pani Panchayat areas also suggest that the low income and low caste farmers were not able to benefit from the creation of WUAs.

Problems Faced by the Farmers

It is revealed from the survey that the sample farmers have been facing many problems, though the major problems reported by the farmers are very few. High cost of maintenance, high cost of electricity, lack of government funding, inadequate field channels, lack of training to staff/members and lack of freedom to determine the water rates have been reported as some of the major problems faced by the sample farmers in the selected areas. Over 40 per cent of the farmers have expressed that high cost of maintenance and high cost of electricity are the major problems that they have been encountering after the formation of WUAs. Inadequate field channels, lack of financial support, lack of training to staff/members have also been reported as the major problems by about 20 to 40 per cent of the sample farmers.

The problems expressed by the canal water users are somewhat similar to the overall results. Over 50 per cent of the farmers have expressed that high cost of maintenance, inadequate field channels, lack of government support (funding) and lack of training to staff/members are the major problems. However, high cost of electricity has not been reported as the major problem by canal water users, because they do not need electricity for irrigation purposes.

The responses of the farmers selected from the Lift Irrigation Users' group are totally different from the one that we have seen with the Canal Water Users' group. About 45 to 60 per cent of the farmers have expressed that lack of government funding and high cost of electricity are the major problems that they have been encountering after the formation of WUAs. Other than these two, the farmers belonging to lift irrigation group have not faced any other major problems. High cost of electricity and lack of financial support (start-up capital) have been reported as the major problems by the farmers selected from Pani Panchayat group.

Assessment on the Success of WUAs

The success of WUAs has also been assessed using the responses of the farmers. Farmers were asked to indicate their perceptions about two points: (a) overall assessment about the success of WUAs and (b) assessment about the financial health of WUAs. On an average, except 10 per cent of the farmers, all other farmers have expressed that overall success of WUAs is either "successful" or "satisfactory". Considerable number of farmers belonging to the tail-end WUAs and poor caste group have expressed that they could not see any big change in water supply and management, even after the formation of WUAs.

The assessment of farmers about the success of WUAs varies with the farmers selected from different WUAs. While about 36 per cent of the farmers selected from Pani Panchayat areas have expressed that the overall success is "poor", the same is not true with the assessment of the farmers selected from canal and lift water users' groups. Since the overall performance of the Pani Panchayat has been deteriorating over the years, the farmers are not willing to admit that their WUA is successful. Over 95 per cent of the farmers selected from canal and lift irrigation water users' groups expressed that the overall performance of WUAs is either "satisfactory" or "successful".

The financial health of WUAs appears to be good from the point of view of the sample farmers. Over 90 per cent of the farmers from canal and lift irrigation water users' group have expressed that the financial health of their respective WUAs is either "successful" or "satisfactory".

However, as expected, about 12 percent of the farmers selected from Pani Panchayat areas have expressed that the financial condition of their institution is "poor".

Policy Recommendations

Though the overall impact of WUAs on different parameters seems to be better, the sample farmers have been facing quite a few problems pertaining to water and other issues. In order to overcome the problems faced by the farmers as well as to sustain the benefits of the users Association, we have presented below some of the policy recommendations.

1. It appears from the responses of the farmers selected from the canal irrigated area that the cost of maintenance of minor, channels and sub-channels are very high. The farmers feel that they may not be able to carry out the operation and maintenance works of the channels without the support of the state agency. Therefore, as suggested by the farmers, the State may continue to share the cost of operation and maintenance so as to reduce the burden as well as sustain the performance of the Water Users' Associations.

2. Though the farmers could see a significant change in the devolution of power after the formation of WUAs, the farmers are not able to decide the timings and quantum of water supply even now. Because of this, the farmers are not able to decide their cropping pattern in advance. Therefore, the State agency may make all arrangements to announce the timings of water supply in advance, before starting of the season.

3. Though Maharashtra State has made significant progress in the formation of WUAs since the mid-nineties, majority of the farmers including the office bearers of WUAs are not fully equipped with the rules and regulations of the formal users' association. Therefore, the State agency can make arrangements to provide training on various aspects of PIM by selecting at least five members from each water users' association. The training on different aspects would help the farmers to solve all the problems pertaining to WUAs easily.

4. Majority of the farmers from canal irrigated area feel that the water rates are very high in Maharashtra State as compared to other neighbouring States. Since the farmers have been paying the water rates in time after the establishment of WUAs, the State government can provide up to 50 per cent of rebate in water rates for those WUAs

who pay the required charges in time – this has been followed in Gujarat State. This would help the Water Users' Association to reduce the water rates and also improve the financial health of the users' institutions.

5. Farmers belonging to both the lift irrigation water users' association and Pani Panchayat system have often been facing erratic electricity supply over the last several years. This often spoils the pumpsets and increases the operation and maintenance expenditures of the pumpsets that they use for lifting water from wells/river/backwater. This also discourages the farmers for cultivating high value crops. Therefore, the State government should ensure quality supply of electricity to the Water Users' Association so as to sustain the performance of the same.

6. The Pani Panchayat model practised in certain part of the State appears to be one of the best systems in protecting the livelihood of the farmers living in the rainfed areas of the State. It has provided substantial benefits to the members of this system over the years. However, our study shows that the current performance of this system is not good mainly because of absence of financial support from State government and high cost of operation and maintenance, especially on account of electricity. Therefore, as suggested by the sample farmers, State government can provide start-up capital as well as subsidy for electricity so as to sustain the Pani Panchayat system.

7. It appears from the responses of the farmers that the Water Users' Associations seem to have not made any big impact on SC/ST and other low income group of farmers. Farmers belonging to this group are mostly poor (small and marginal) and they may not able to make use of the benefits coming through Water Users' Association, without the interference of the State agency. Therefore, the State can make adequate arrangements to see that the benefits of WUAs are distributed to these groups of farmers so as to achieve the "inclusive growth" envisaged in the approach paper of the 11th Five Year Plan.

Conclusion

The results of the study suggest that the model of participatory irrigation management implemented in the canal irrigated areas of Maharashtra State seems to be working reasonably well at this point of time. Farmers' responses clearly reveal positive changes in the devolution of power from the State agency to WUAs particularly in cases like pricing of water, collection of dues, operation and maintenance of canal, allocation of water for farmers and for different crops, etc. Farmers by and large seem to be happy because they are able to manage almost all the water related problems within the villages after the formation of WUAs. The formation of WUAs has also made better impact on cropping pattern, expansion of irrigated area, input use and productivity of crops as well as on the overall economic development of the selected regions. These positive benefits from WUAs would not have come without the effort of the State government, which has been taking serious effort to implement the participatory irrigation management in phased manner since mid-nineties.

Unlike some of the Indian States, Maharashtra State has followed "bottom up approach", wherein the State agency has made all out arrangements to explain the benefits of PIM through various training and other extension programmes for the farmers and encouraged them to form WUAs on their own, before making a formal announcement for handing over the system management to farmers. This is also very evident from the strategy of the government of Maharashtra that has made policy decision to handover the irrigation management system in 2001 itself, but it could bring an Act to legalise the same only in May 2005. In the mean time, the State agency could work effectively to strengthen the software aspects of irrigation management. This is something different from Andhra Pradesh's PIM model, wherein "big bang" approach was followed without providing basic training and exposure to farmers on PIM and its advantages. The initial experience of the State suggests that the participatory irrigation management especially in canal irrigated area can be made successful, if it is introduced in a phased manner along with the required basic training to farmers.

Though the overall performance of WUAs functioning in the canal irrigated area is reasonably good, one cannot immediately take this as a strong endorsement for the participatory irrigation management. Most of the WUAs are young and still get some kind of support from the State agency. We do not know what would be the status of WUAs if the State completely withdraws all the support that is given to WUAs. Similarly, the overall performance of WUAs may descend, if water is not supplied in time to WUAs by the State agency. Moreover, the problems faced by the farmers pertaining to Water Users' Associations are expected to change from time to time because of changing nature of institutional and other factors. Farmers may not perceive operation and maintenance of the minor or channels as a serious problem at this moment, but it may become a serious problem after two or three years. There is no guarantee that WUAs alone would be able to manage the operation and maintenance works of the system without financial support from the State. Therefore, studies need to be carried out periodically (once in three or five years) to understand the performance as well as to identify the problems of water users so as to sustain the benefits of the water users' association. One can hope that WUAs would be able to carry out its specified objectives and duties as long as it is able to manage smooth relationship with the State agency.

ANNEXURE

MAHARASHTRA MANAGEMENT OF IRRIGATION SYSTEMS BY FARMERS ACT, 2005

MAHARASHTRA ACT NO XXIII OF 2005

To provide for Management of Irrigation Systems by Farmers' and for matters connected therewith or incidental thereto

WHEREAS the National Water Policy, (2002) lays down that efforts should be made to progressively involve farmers, in the various aspects of management of irrigation systems;

AND WHEREAS the Maharashtra Water and Irrigation Commission (1999) has recommended that statutory provisions may be made for management by farmers, of irrigation systems by providing water from Public Canal System to Water Users' Associations on volumetric basis;

AND WHEREAS it is decided by the State Government to bridge the gap between the irrigation potential created and its actual utilisation and to optimise the benefits by ensuring proper use of surface and groundwater by increased efficiency in distribution, delivery, application and drainage of irrigation systems and for achieving this objective to give statutory recognition to the constitution and operation of Water Users' Associations so as to enable the farmers to act collectively to improve the productivity of agriculture;

AND WHEREAS both Houses of the State Legislature were not in session;

AND WHEREAS the Governor of Maharashtra was satisfied that circumstances existed which rendered it necessary for him to take immediate action, to make necessary provisions, for the purposes aforesaid; and, therefore, promulgated the Maharashtra Management of Irrigation Systems by Farmers Act, 2004, on the 18th August, 2004:

AND WHEREAS it is expedient to replace the said Act by an Act of the State Legislature : it is hereby enacted in the Fifty-sixth Year of the Republic of India, as follows :-

Chapter I

PRELIMINARY

1. (1) This Act may be called the Maharashtra Management of Irrigation Systems by Farmers Act, 2005.

(2) It extends to the whole of the State of Maharashtra.

Mah. XXXVIII of 1976

(3) Save as otherwise expressly provided in this Act, it shall come into force on such date as the State Government, may by notification in the Official Gazette, appoint and different dates may be appointed for different areas and for different provisions :

Provided that, the areas in respect of which the notification under section 5 has not been issued, it shall not come into force in such area and the provisions of the Maharashtra Irrigation Act, 1976 shall continue to apply in the area.

Definitions

2. In this Act, unless the context otherwise requires,-

(I) (a) "adequate repairs and renovations", in relation to handing over of Canal System to Water Users' Association, means those repairs and renovations which are essential for passing designated discharge, control and measuring of water, regulation of water level and convey water to all fields, by flow under gravity, in the area of operation of Water Users' Association;

(b) "Agreement" means an agreement between the Water Users' Association and the Upper Level Water Users' Association or Canal Officer, as the case may be, for the supply of water on volumetric basis for irrigation and handing over of physical system for operation, maintenance and management;

(c) "Applicable Water Entitlement (Water Quota)" in relation to a reservoir supported canal system, means the total volume of water guaranteed to be supplied to a Water Users' Association at the agreed point of supply, in any year arrived at according to the rules made under this Act, by the Appropriate Authority and in the case of Minor Irrigation Projects having only one Water Users' Association, available water for Irrigation at Canal Head;

(d) "Appropriate Authority" means the Appropriate Authority as defined in clause (2) of section 2 of the Maharashtra Irrigation Act, 1976 and also includes Maharashtra Water Resources Regulatory Authority and Irrigation Development Corporations established by the State Government by Acts of the State Legislature ;

(e) "area of Operation" in relation to Water Users' Association means the area notified under section 5 in respect of such Association;

(f) "Canal Officer" means the Canal Officer appointed under clause (4) of section 2 of the Maharashtra Irrigation Act, 1976, who is duly empowered by the State Government, by notification in the Official Gazette, for the purposes of exercising the powers and discharging the duties under this Act;

(g) "Canal system" means all Canals and includes all lands commanded thereby;

(h) "Command Area" in relation to flow and lift irrigation on a canal, means the land under irrigable command of a canal, bound by hydraulic or other boundaries for which a Water Users' Association is formed;

(i) "Competent Authority" means the Authority appointed under Section 38;

(j) "Culturable Command Area" means the area, under an Irrigation Project, which is culturable and can get irrigation facility through Canal System;

(k) "Deficit Year" in relation to a diversion system or reservoir based canal system, means a year in which the water availability at the head works is less than that in a Normal Year;

(l) "Designated Discharge" means designed discharge or a revised designed discharge which the Canal System can carry after its rehabilitation or renovation, as the case may be;

(m) "Entitlement" means bulk water use entitlement or individual water use entitlement, as determined by Appropriate Authority from time to time;

(n) "General Body" in relation to a Water Users' Association means the body of all members of such Association;

(o) "Hydraulic Basis" means the basis for identifying an area of operation of Water Users' Association served by one or more hydraulic structures such as Head Regulator of Canal, Distributary, Minor, Sub-Minor, Field Channel Lift Irrigation Scheme or irrigation through lifting water by pumps and similar regulatory structures ;

(p) "Irrigation Development Corporation" means an Irrigation Development Corporation established by the State Government by an Act of the State Legislature;

(q) “Lands under Management of Irrigation Systems by Farmers” means lands under irrigable command of a canal where the management of irrigation water distribution with all its related aspects including responsibility of maintenance is handed over to Water Users' Association;

(r) "Lift Irrigation" or “Lift Irrigation Scheme” in relation to Lift Irrigation Water User’s Association means all Lift Irrigation Schemes, wherein Government permission is required for lifting water, but excluding the Lift Irrigation Schemes on private open wells and bore wells:

(s) “Lift Irrigation Water Users' Association (LIWUA)”, in relation to MISF, means a association constituted for the purpose of lift irrigation, registered in the prescribed manner as provided in section 43;

(t) “Maharashtra Water Resources Regulatory Authority (MWRRA)” means the authority established by the State Government by an Act of the State Legislature for regulation of water resources in the State;

(u) “Major Irrigation Project”, Medium Irrigation Project” and “Minor Irrigation Project” means project having Culturable command area exceeding ten thousand hectares, exceeding two thousand hectares but not exceeding ten thousand hectares, and not exceeding two thousand hectares, respectively:

(v) “Management of Irrigation Systems by Farmers (MISF)” means involvement of irrigation users in all aspects of irrigation, including planning, designing, constructing, supervising, financing, operating, managing, maintaining, monitoring and evaluating of the Canal System;

(w) “Member”, means a member of a Water Users’ Association who is either a land holder or occupier of the delineated area of the Water

Users' Association and includes Water Users' Association of another upper level association constituted under the provisions of this Act ;

(x) "Non-Irrigation Water Supply" means water supply from an irrigation project for purposes other than irrigation and includes water supply for the purposes such as domestic, industrial, power generation, fisheries and as may be specified by the Appropriate Authority from time to time;

(y) "Normal Year" in relation to a diversion system or a reservoir based canal system means a year in which the water availability at the headworks (including the water that is allowed to be lifted from the reservoir and losses therefrom) is as envisaged in the Irrigation Project Report or as revised subsequently by the State Government;

(z) "prescribed" means prescribed by Rules made under this Act.

(aa) "Prescribed Unit Water Use Entitlement" in relation to a reservoir based canal system means the total volume of water that is available in an irrigation year or season per hectare of Culturable Command Area at the point of supply to Water Users' Association, arrived at after considering prior sanction for Non irrigation purposes and all the losses and gains from the canal system in a Normal Year;

(ab) "Prescribed Unit Water Allowance" in relation to a canal system, which is not reservoir supported, means the discharge allocable per hectare of the Culturable Command Area, measured at the point of supply to a Command Area;

(ac) "Previous Dues", in relation to assessment and recovery of water charges, means the outstanding dues of members of Water Users' Association pertaining to the period prior to the date of handing over the management of irrigation systems ;

(ad) "Reservoir based canal system" means a canal system where certain per cent. of the volume of water as specified by the Appropriate Authority is stored making the availability of water predictable for use ;

(ac) "rules" means the rules made under this Act;

(af) "Rotational Water Supply (RWS)" means a system of water distribution in which water is supplied to water users by rotation, according to an approved schedule indicating the day, duration and time of supply;

(ag) "Sanctioned Water Use Entitlement" in relation to an Agreement with Water Users' Association on a reservoir based canal system means, the total volume of water guaranteed to be supplied to the Water Users' Association, at the agreed point of supply, in different seasons, in a Normal Year;

(ah) "State Government" or "Government" means the Government of Maharashtra;

(ai) "Surplus Year" in relation to a diversion system or reservoir based canal system means a year in which the water availability at the headworks is more than that in the Normal Year;

(aj) "Water Allowance" means the discharge allotted at the point of supply to the Water Users' Association;

(ak) "Water Users" means and includes any individual or body corporate or an association, using water from a Government sources of irrigation;

(al) "Water Users' Association" in relation to Management of Irrigation systems by Farmers means all types of Water User's Association at all levels of canal system for flow as well as lift irrigation;

(am) "Water Users' Association at Minor Level" means the primary unit of Water Users' Association;

(II) Words and expressions used in this Act, but not defined herein, shall have their respective meaning as assigned to them in the Maharashtra Irrigation Act, 1976.

Chapter – II

WATER USERS' ASSOCIATION AND DELINEATION OF COMMAND AREA THEREOF

Water Users' Association to be body Corporate

3. Every Water Users' Association (WUA) constituted under this Act shall be a body corporate with a distinct name having perpetual succession and a common seal and shall have the power to acquire, hold and dispose off property and to enter into contracts and of doing all things necessary, proper or expedient for the purposes for which it is constituted and may sue or be sued in its corporate name :

Provided that, no Water Users' Association shall have the power to alienate in any manner, any property vested in it by the Government for the operation and management.

Objects of Water Users Association

4. (1) The objects of a Water Users' Association shall be to ,-

(i) promote and secure equitable distribution of water amongst its members;

(ii) maintain adequately the irrigation systems; and ensure efficient, economical and equitable distribution and utilisation of water to optimise agricultural production;

(iii) protect the environment;

(iv) ensure ecological balance

(v) Actively involve the members inculcating amongst them a sense of ownership of the irrigation system; and

(vi) Safeguard and promote the common interests of its members pertaining to irrigation and agriculture in the area of operation.

(2) The Association may also engage into any activity of common interest of the members in the Command Area related to irrigation and agriculture, such as introduction of Drip and Sprinkler system for optimising the use of water; developing farm ponds and community projects for exploiting groundwater; procurement and distribution of seeds, fertilisers and pesticides; procurement and renting of agricultural implements; marketing and processing agricultural produce from the Command Area and supplementary business like dairy and fisheries.

Delineation of areas as Areas under Management of Irrigation System by Farmers

5. For the purpose of this Act, Appropriate Authority may, by notification in the Official Gazette, delineate, on hydraulic basis and having regard to the administrative convenience, the command area of an irrigation project into an area under the Management of Irrigation System by Farmers, for which there shall be constituted Water Users' Association under this Act and may by like notification from time to time, alter their limits, by– (i) amalgamating or dividing the Water Users' Associations; (ii) increasing an area of any Water Users' Association; (iii) reducing an area of any Water Users' Association;

Provided that, no such notification making amalgamation, division, or in any way altering the boundaries of the Water Users' Association shall be issued unless a reasonable opportunity, as prescribed, is given to the Water Users' Association and holders or occupiers of the lands likely to be affected thereby.

Provided further that, no Water User's Association and land holder or occupier shall be eligible to obtain membership of any other Water

Users' Association unless 'No dues Certificate' of the previous Water Users' Association, is produced.

Delineation of Command Area of Water User's Association at minor level

6. (1) A Canal Officer not below the rank of an Executive Engineer, duly empowered in this behalf, may, by notification in Official Gazette and in accordance with the rules, on hydraulic basis and having regard to the administrative convenience delineate command area of Water Users' Association at Minor Level separately within lands under irrigable command and declare it to be an area of operation of Water Users' Association at Minor Level for the purposes of this Act. The area of operation so declared may include both, the flow and lift irrigation.

(2) The notification under sub-section (1) shall contain, (a) a certified copy of updated map of area of operation of Water Users' Association showing prescribed details necessary for Operation and Management of such Association ; and (b) a certified copy of updated list of landholders and occupiers of land with prescribed details within the area of operation of Water Users' Association at Minor Level.

(3) The notification published under sub-section (1) may be given wide publicity as prescribed and may be published in such other manner in or in the vicinity of such area of operation of Water Users' Association at Minor level as the Canal Officer may think fit.

(4) Any person desirous of obtaining a copy of the notification under sub- section (1) may, on an application being made in that behalf and on payment of such fees as may be prescribed, obtain the same.

(5) Any person aggrieved by the notification issued under subsection (1) may, within thirty days from the date of publication of such notification in the Official Gazette, file an appeal before the Canal Officer:

Provided that, no Canal Officer below the rank of a Superintending Engineer shall be notified as the appellate officer by the Government.

(6) The Appellate Officer may pass such order in the appeal as he deems fit, and may pass order modifying the notification. On passing such modifying order, such notification shall stand modified to the extent and from the date as specified in the order; Provided that, no order of modification of any such notification or part thereof, affecting any person shall be made without giving such person a reasonable opportunity of being heard.

Supply of water to land under Management of Irrigation

7. On delineation of Command area of Water Users' Association, by notification in the Official Gazette under section 6, no water shall be supplied by the appropriate Authority to the System by Farmer's. Individual holder or occupier: and the system of supply of water through Water Users' Association shall be binding on all holders and occupiers of such lands.

Constitution of Water Users' Association at Minor Level

8. (1) When an area of operation of a Water Users' Association at Minor Level has been delineated under section 6, the holders and occupiers of the land so delineated shall form a Water Users' Association. Such Water Users' Association shall be registered in the prescribed manner.

Provided that, where no such Water Users' Association is constituted, the Appropriate Authority may, entrust the functions of such Water Users' Association to Government agency or Cooperative societies or any other Water Users' Association in the manner prescribed, till such Association is constituted and functional.

(2) All the holders and occupiers in delineated land of Water Users' Association at Minor Level shall be deemed to be the members of Water Users' Association at Minor Level, and shall constitute the General Body of the Water Users' Association and shall have right to vote as prescribed.

Managing Committee of Water Users' Association at Minor Level and Election of its Directors and Chairperson

9. (1) There shall be a Managing Committee for every Water Users' Association at Minor Level, consisting of such number of members including Chairperson and Directors as may be prescribed. Such Association shall have a Secretary having prescribed qualifications and the pay and allowances of such person shall be fixed by the respective Water Users' Association, in the prescribed manner.

(2) The Canal Officer, duly empowered in this behalf shall, make arrangements for the election of Directors of the Managing Committee of the Water Users' Association at Minor Level by direct election from amongst members of such Water Users Association within such time and in manner as may be prescribed.

(3) Adequate representation shall be given on the Managing

Committee to the holders or occupiers of the land from Head, Middle and Tail reaches of the area of operation of Water Users' Association; and Women members in such manner as may be prescribed.

(4) Such Canal Officer shall make arrangements for the election of the Chairperson of the Water Users' Association at Minor Level in prescribed manner. The Directors of the Managing Committee shall elect the Chairperson in the first meeting of the Managing Committee convened and presided by the Canal Officer pending the election of Chairperson.

(5) The tenure of the post of Chairperson, its rotation amongst the Managing Committee Directors representing Head, Middle and Tail reaches of the area of operation of Water User's Association and women Directors shall be such as may be prescribed: Provided that, where a land holder or occupier holds or occupies a land in the area of operation of more than one Water User's Association, such person shall be eligible to take part in the election for the Directorship of the Managing Committee of only one Water User's Association opted by him in the prescribed manner.

(6) The Canal Officer may, for the reasons to be recorded in writing, postpone the election for a period up to one year.

(7) The Directors of Managing Committee shall, unless recalled earlier under section 19, hold office for a period of six years from the date of the constitution of the said Managing Committee.

(8) The Managing Committee shall exercise such powers, discharge such duties and perform such functions of the Water Users' Association, as provided in this Act and as prescribed.

(9) Notwithstanding anything contained in sub-section (2), the Canal Officer, duly empowered in this behalf, may pending election and constitution of the Managing Committee appoint an Ad-hoc Management Committee in the prescribed manner.

Delineation of Area of Operation of Distributary Level Association

10. (1) A Canal Officer not below the rank of an Executive Engineer, duly empowered in this behalf, may by notification in Official Gazette and in accordance with the Rules delineate, on hydraulic basis and having regard to the administrative convenience ,command area of Distributary Level Association (DLA) within Lands Under Irrigable Command and declare it to be an area of operation of distributary level Association for the purposes of this Act. The area so declared may include both, the flow and lift irrigation.

(2) The notification under sub-section (1) shall contain, (a) a certified copy of updated map of area of operation of Distributary Level Association showing prescribed details necessary for Operation and Management of Canal System under Distributary Level Association; (b) a certified copy of the updated list of Water Users' Associations at Minor level included in the Distributary Level Association with prescribed details.

(3) The provisions of sub-sections (3), (4), (5) and (6) of section 6 shall, Mutatis Mutandis, apply to the notification under subsection (1).

Constitution of Distributary Level Association

11. (1) In the area delineated as an area of operation of Distributary Level Association under Section 10, a Distributary Level Association shall be constituted by the Water Users' Associations at Minor Level. Such Association shall be registered in the prescribed manner.

(2) All the Water Users' Associations at Minor Level, represented by the members of Managing Committees of the Water Users' Associations at Minor Level, in the command area of Distributary Level Association shall constitute the General Body of the Distributary Level Association and shall have right to vote and elect the Managing Committee of the Distributary Level Association, in such manner as may be prescribed.

(3) After the operation and maintenance of the delineated area is handed over to such Distributary Level Association, water supply to water users in such delineated area shall be through the respective Distributary Level Association.

Managing Committee of Distributary Level Association and Election of its Members and Chairperson

12. (1) There shall be a Managing Committee for every Distributary Level Association (DLA) consisting of such number of members, including Chairperson and Directors as may be prescribed. Such Association shall have a Secretary, having prescribed qualifications: and the pay and allowances of such person shall be fixed by the respective Water Users' Association, in the prescribed manner.

(2) The Canal Officer, duly empowered in this behalf, shall make arrangements for the election of Directors of the Managing Committee of Distributary Level Association by election from amongst members of General Body of Minor Level Association, within such time and in such manner, as may be prescribed.

(3) Adequate representation shall be given on the Managing Committee of Distributrary Level Association to the Water Users' Associations from Head, Middle and Tail reaches of the area of operation of Distributary Level Association; and women Directors, in the prescribed manner.

(4) Such Canal Officer shall also make arrangements for the election of the Chairperson of the Distributary Level Association in the manner as may be prescribed. The Directors of Managing Committee shall elect the Chairperson in the first meeting of the Managing Committee convened and presided over by the Canal Officer, pending election of the Chairperson.

(5) The tenure of the post of Chairperson, and its rotation amongst the Managing Committee Directors representing Head, Middle and Tail reaches of the area of operation of Distributary Level Association and women Directors shall be such as may be prescribed.

(6) The Canal Officer may, for the reasons to be recorded in writing, postpone the election up to one year.

(7) The Members of Managing Committee shall, unless recalled earlier under section 19, shall hold office for a period of six years from the date of the constitution of the said Managing Committee.

(8) The Managing Committee shall exercise such powers, discharge such duties and perform the functions of the Distributary Level Association as may be provided in this Act and as prescribed.

(9) Notwithstanding anything contained in sub-section (2), the Canal Officer, duly empowered in this behalf pending election and constitution of Managing Committee may, appoint an Ad-hoc Management Committee in the prescribed manner.

Delineation of area of Operation of Canal Level Association

13. (1) A Canal Officer not below the rank of an Executive Engineer, duly empowered in this behalf, may, by notification in the Official Gazette, and in accordance with the Rules delineate on hydraulic basis and having regard to administrative convenience, command area of canal level Association (C.L.A.) within lands under irrigable command and declare it to be an area of operation of Canal Level Association for the purposes of this Act. The area of operation so declared may include both, flow and lift irrigation.

(2) The notification as per sub-section (1) shall contain ---(a) a certified copy of updated map of area of operation of Canal Level

Association showing prescribed details necessary for Operation and Management; (b) a certified copy of the updated list with prescribed details of Distributary Level Associations included in the Canal Level Association.

(3) The provisions of sub-sections (3), (4), (5) and (6) of section 6 shall, Mutatis Mutandis, apply to the notification under sub-section (1).

Constitution of Canal Level Association

14. (1) In the area delineated as an area of operation of Canal Level Association (CLA) under section 13, a Canal Level Association shall be constituted by the Distributary Level Associations. Such Association shall be registered in the prescribed manner.

(2) All the Distributary Level Associations, represented by the Directors of the Managing Committees of the Distributary Level Associations, under the command of Canal Level Association shall constitute the General Body of Canal Level Association and have right to vote and elect the Managing Committee of the Canal Level Association in such manner as may be prescribed.

(3) After the operation and maintenance of the delineated area is handed over to Canal Level Association, water supply to water users in such delineated area shall be through the respective Canal Level Association.

Managing Committee of Canal Level Association and Election of its Directors and Chairperson

15. (1) There shall be a Managing Committee for every Canal Level Association (CLA) consisting of such number of members, including Chairperson and Directors as may be prescribed. Such Association shall have a Secretary having prescribed qualifications: and the pay and allowances of such person shall be fixed by the respective Water Users' Association in the prescribed manner.

(2) The Canal Officer, duly empowered in this behalf, shall make arrangements for the election of Directors of the Managing Committee of Canal Level Association by election from among members of General Body of Canal Level Association, within such time and in such manner as may be prescribed.

(3) Adequate representation shall be given on the Managing Committee of Canal Level Association to the Distributary Level Associations from Head, Middle and Tail reaches of the area of operation

of Canal Level Associations; and Women in such prescribed manner.

(4) Such Canal Officer shall also cause arrangements for the election of the Chairperson of the Canal Level Association in the manner as may be prescribed. The Directors of the Managing Committee shall elect the Chairperson in the first meeting of the Managing Committee convened and presided over by the Canal Officer, pending election of the Chairperson.

(5) The tenure of the post of Chairperson, its rotation amongst the Managing Committee Directors representing Head, Middle and Tail reaches of the area of operation of Canal Level Associations and women members shall be such as may be prescribed.

(6) The Canal Officer may, for the reasons to be recorded in writing, postpone the election up to one year.

(7) The Directors of Managing Committee shall, unless recalled earlier under section 19, hold office for a period of six years from the date of the constitution of the said Managing Committee.

(8) The Managing Committee shall exercise such powers, discharge such duties and perform such functions of the Canal Level Association as may be provided in this Act and as prescribed.

(9) Notwithstanding anything contained in this sub-section (2), the Canal Officer, duly empowered in this behalf, may, pending election and constitution of Managing Committee appoint an Ad-hoc Managing Committee in the manner prescribed.

Delineation of area of Operation of Project Level Association

16. (1) A Canal Officer not below the rank of Superintending Engineer, duly empowered in this behalf, may by notification in Official Gazette and in accordance with the rules prescribed . delineate lands under irrigable command of the Project and declare it to be an area of operation of Project Level Association (PLA) for the purpose of this Act. The area of operation so declared may include both flow and lift irrigation.

(2) The notification under sub-section (1) shall contain ---(a) a certified copy of updated map of area of operation of Project Level Association showing areas of operation of all Water Users' Associations at different levels under the Project and other prescribed details necessary for Operation and Management; (b) a certified copy of the updated list with prescribed details of all Water Users' Associations at different levels included in the Project Level Association.

(3) The provisions of sub-sections (3), (4), (5) and (6) of section 6 shall, Mutatis Mutandis, apply to the notification under sub-section (1).

(6) Notwithstanding anything contained in this section, in the case of Minor Irrigation Project, notified Executive Engineer shall delineate the area of operation of Project Level Association and the appellate officer shall be the notified Superintending Engineer.

Constitution of Project Level Association

17. (1) In the area delineated as an area of operation of Project Level Association (PLA) under Section 16, a Project Level Association shall be constituted by the Canal Level Associations functioning within the project. Such Association shall be registered in the prescribed manner.

(2) All Canal Level Associations and Lift Irrigation Water Users' Associations on reservoir within the project, represented by Directors of their respective Managing Committees, shall constitute the General Body of Project Level Association and shall have right to vote and elect the Managing Committee of the Project Level Association in such manner as may be prescribed.

(3) After the operation and maintenance of the delineated area is handed over to Project Level Association, water supply to water users in such delineated area shall be through the respective Project Level Association.

(4) Notwithstanding anything contained in sub-section (1), the State Government may, having regard the administrative convenience, for reasons to be recorded in writing, authorise an association at any level, to function as a Project Level Association in respect of a project. On such authorization, the provisions of sub-section (2) and other provisions of the Act applicable to the Project Level Association shall, Mutatis Mutandis, apply to such association.

Managing Committee of Project Level Association and Election of its Directors and Chairperson

18. (1) There shall be a Managing Committee for every Project Level Association (PLA) consisting of such number of Members including Chairperson and Director as may be prescribed. Such Association shall have a Secretary having prescribed qualifications: and the pay and allowances of such person shall be fixed by the respective Water Users' Association in the prescribed manner.

(2) The Canal Officer, duly empowered in this behalf, shall make

arrangements for the election of Directors of the Managing Committee of Project Level Association by election from amongst members of General Body of Project Level Association, within such time and in such manner, as may be prescribed.

(3) Adequate representation shall be given on the Managing Committee of Project Level Association to member-Water Users' Associations from Head, Middle and Tail reaches of the area of operation of Project Level Associations and Women members, in the prescribed manner.

(4) Such Canal Officer shall also make arrangements for the election of the Chairperson of the Project Level Associations in the manner as may be prescribed. The Directors of Managing Committee shall elect the Chairperson in the First Meeting of the Managing Committee convened and presided over by the Canal Officer pending election of the Chairperson.

(5) The tenure of the post of Chairperson, its rotation amongst the Managing Committee Directors representing Head, Middle and Tail reaches of the area of operation of Project Level Association and Women Directors shall be such as may be prescribed.

(6) The Canal Officer may, for the reasons to be recorded in writing, postpone the election up to one year.

(7) The Directors of Managing Committee, unless recalled earlier under section 19, shall hold office for a period of six years from the date of the constitution of the said Managing Committee.

(8) The Managing Committee shall exercise such powers, discharge such duties and perform such functions of the Project Level Association as may be provided in this Act as prescribed.

(9) Notwithstanding anything contained in sub-section (2), the Canal Officer, duly empowered in this behalf, may, pending election and constitution of the Managing Committee appoint an Ad-hoc Managing Committee in the manner prescribed, till the elected Managing Committee comes into existence.

Procedure for Recall

19. (1) A motion for recall of a Director of a Managing Committee of a Water Users' Association at any level may be made by giving a notice, in writing in the prescribed form, signed by not less than one-half of the total number of members of the concerned Water Users' Association, entitled to vote;

Provided that, no such motion against any office bearer shall be allowed within two years from the date of assumption of office by such office bearer.

(2) If the motion referred to in sub-section (1), is carried with the support of two-third of the Members present and voting at a meeting of the General Body specially convened for the purpose, the Canal Officer shall, by order, remove the office bearer, against whom such motion is passed and the resultant vacancy shall be filled in prescribed manner.

Constitution of Subcommittees of Water User's Association Agreement

20. The Managing Committee of a Water User's Association may constitute sub-committees to carry out all or any of the functions vested in such Water User's Association under this Act.

21. (1) There shall be an agreement between Water Users' Association (WUA) and the upper level Water Users' Association or the Canal Officer, of the rank of Superintending Engineer or Executive Engineer as the case may be, within three months of the constitution of the concerned Water Users' Association. Such agreement shall contain the contents mentioned in section 29.

(2) Such agreement shall be valid for such period as may be prescribed.

Joint Inspection, Repairs, Rehabilitation and Handing over of Management of Command Area to Minor Level Water Users' Association

22.(1) After the Water Users' Association at a minor level has been duly constituted and the agreement between Water Users' Association and the Canal Officer, as the case may be, has been entered into, there shall be held a joint inspection of the canal system under the said Agreement within a period of three months from the date of Agreement. The entire canal system within the command area of Water User's Association, shall be jointly inspected by the Executive Engineer or his representative and the office-bearers of the Water Users' Association and the works pertaining to the systems shall be classified into two separate lists i.e. Priority-I and Priority-II.

(2) The Priority-I List shall include the mandatory nature of works, which are absolutely essential for passing of designated discharge, control and measuring and conveyance of water, by flow under gravity

in the area of operation of WUAs. The Priority II List shall include the works other than those mentioned in Priority-I which although necessary for the efficient functioning of WUAs, can be taken up after the Priority-I works.

(3) The findings of the joint inspection shall be recorded in duplicate, signed by both the minor level Water Users' Association and the Canal Officer and one copy thereof shall be retained by each party.

(4) Within one month of joint inspection, a list of items of works which need to be carried out for rehabilitation of the canal system, based on the findings of such joint inspection and in accordance with the directives and orders of the Appropriate Authority, and agreed to between the Canal Officer and the Water Users' Association, shall be prepared.

(5) On preparation of such list of items of works under subsection (4), the Canal Officer not below the rank of an Executive Engineer duly empowered in this behalf, shall inform the Water Users' Association, the schedule of completion of repairs and renovation, if any required, to ensure that water can be delivered to all the command area of the canal system, in the designeted and specified period. The schedule of completion for the works of Priority-I, shall not extend beyond a period of twelve months, from the date of joint inspection, except with the consent of the concerned Water Users' Association. After completion of Priority-I works, the system will be tested and handed over to WUA. If the Water Users' Association fails to take over the system, within one month of such testing, it will stand handed over to the Water Users' Associations. The participation of Water Users' Association in repairs and renovation works shall be as prescribed. After such handing over of the system, Priority-II works shall be completed within a period not exceeding eighteen months. The Canal Officer shall have power to extend such period in exceptional circumstances.

(6) There shall be annexed to such document, the copies of the joint inspection findings, the agreed schedule of completion of the repairs and renovation, together with the list of items of such repairs and renovation agreed to by both the parties

(7) If any dispute between Water Users' Association and the Canal Officer arises regarding anything contained in an agreement executed in accordance with sub-section (1) , the same shall be resolved by the concerned Superintending Engineer and his decision in the matter shall be final.

Installation of Measuring Device

23. (1) For every area of operation delineated under this Act or where a Water Users' Association for flow irrigation has been duly constituted under this Act, it shall be the duty of the Canal Officer to provide a proper measuring device or devices on the canal at the point of supply to Water Users' Association and ensure its proper working from time to time.

(2) The accurate flow measurement, the form of record in which it shall be entered into and periodic evaluation thereof; as well as the mode of ascertaining the volume of water for a period in which measuring device is out of order, shall be such as may be prescribed.

Freedom of Cropping

24. The Water Users' Association shall have the freedom of growing different crops within Applicable Water Entitlement or Water Allowance, as the case may be, subject to any restrictions mentioned in Agreement.

Freedom of Recycling Irrigation Water and use of Groundwater

25. (1) The Water Users' Association in Farmers' Management of Irrigation System areas shall have the freedom of recycling of the water, at its own cost, for irrigation.

Provided that, there shall not be any extra charge to be paid to the Appropriate Authority for such recycling of water.

(2) The Water Users' Associations shall also have the freedom of using groundwater in their command area, in the prescribed manner, in conjunction with canal water or otherwise:

Provided that, there shall not be any extra charge to be paid to the Appropriate Authority for use of groundwater.

Mah. XXXVIII of 1976

(3) In order to ensure equity of water supply within project and river basin and to protect environment, the Appropriate Authority may prohibit the formation of obstructions to drainage within certain limits by taking recourse to the provisions of sections 19, 20 and 21 of Maharashtra Irrigation Act, 1976.

Modes and Rates for Supply of Water to Water Users' Association

26. (1) Water from the canal system shall be supplied to the Water Users' Associations (WUAs) at various levels, from tail to head on bulk

basis measured volumetrically as per their water entitlements by Canal Officer or upper level Water Users' Association, as the case may be.

(2) The rates for supply of water to a Water Users' Association shall be on the volumetric basis measured at the point of supply.

(3) The Appropriate Authority shall have the power to levy the minimum charges as prescribed to Water Users' Association if water is not demanded or used for irrigation by Water Users' Association in a season as per the Applicable Entitlement.

(4) The rates for supply of water under sub-section (2) and the minimum charges under sub-section (3) shall be such as may be prescribed.

Powers of Water Users' Association to Charge for Supply of Water to members

27. (1) The Water Users' Association shall have powers and responsibility to charge to its members, water rates as may be approved by the General Body of the Water Users' Association.

(2) Water Users' Association shall have the power to levy the minimum charges for the land for which water is not demanded or used for irrigation by members :

Provided that, no such minimum charge shall be levied if the water is not available as per the sanctioned Water use entitlement.

(3) The Water Users' Association shall also have the power to levy the water charges for use of recycled water or ground water by members.

(4) The Water Users' Association shall be entitled to recover the previous dues from its members in the prescribed manner.

Supply of water as per Entitlement

28. (1) It shall be the responsibility of the Appropriate Authority to supply water as per the Applicable Water Use Entitlement, in the prescribed manner to the Water Users' Associations on a bulk basis measured volumetrically.

(2) It shall be the responsibility of the Water Users' Association to supply water equitably in its area of operation as per Applicable Water Entitlement of each member.

Contents of Agreement

29. (1) Supply of water for irrigation to any Water Users' Association

at any level within the area of Management of Irrigation Systems by Farmers shall be in accordance with the Agreement to be signed by the Chairperson of Water Users' Association and the Chairperson of upper level of Association or the Canal Officer duly empowered in this behalf as the case may be. A note of such agreement shall be kept with the Water Resources Department in the prescribed manner.

(2) The agreement shall, inter alia, contain the provisions for following:-

(i) Objectives of the agreement;

(ii) Water use Entitlement;

(iii) Water rate and assessment on volumetric basis;

(iv) Rights of member of the Water Users' Association;

(v) Rights of Water Users' Association;

(vi) Recovery of water charges;

(vii) Previous dues;

(viii) Maintenance and Repairs of Canal System;

(ix) Maintenance and Repairs of field channels and field drains;

(x) Special incentives, if any, given by Appropriate Authority from time to time;

(xi) Resolution of conflicts;

(xii) Period of Agreement;

(xiii) Termination or Revision of Agreement;

(xiv)Joint inspections, rehabilitation, schedule of balance work and handing over, etc;

(xv) Compensation;

(xvi) Penalties;

(xvii) Technical guidance and training;

(xviii) All other matters as prescribed which need to be provided for carrying out the purposes of this Act, in the Agreement under this Act.

Office Bearers of Water Users' Association and their Powers

30. In order to carry out functions of Water Users' Association and to achieve the objectives, of this Act, the office-bearers of Water Users' Association and any person acting under the general or specific

instructions of such office-bearers shall have the powers as may be prescribed.

Power to Remove Encroachments

31. A Water Users' Association may remove encroachments from property attached to the canal system within its area of operation in accordance with such procedure as may be prescribed.

Resignation

32. (1) A Director of the Managing Committee of a Water Users' Association, at any level may resign his office by a letter sent by registered post or tendered in person to the Chairperson of the Managing Committee concerned.

(2) The Chairperson of the Managing Committee of a Water Users Association at Minor Level may resign his office by a letter sent or tendered in person to the Chairperson of the Managing Committee of the immediate next upper level Association concerned.

Provided that, the Chairman of the Managing committee of the Project level Association may resign his office by a letter sent by the registered post or tendered in person to the concerned Canal Officer.

Explanation.– In cases, where the Managing Committee of the upper level Association at any level does not exist, then the resignation of the Chairperson of Managing Committee shall be sent by registered post or tendered in person to the chairperson of the Managing committee of the next upper level Association or the concerned Canal Officer, as the case may be.

(3) The resignation letters received under sub-sections (1) and (2) shall immediately be forwarded to the Managing Committee of the Water Users' Association concerned for acceptance or otherwise and a copy of such resignation letter shall also be forwarded to the concerned Canal Officer.

(4) Every such resignation shall take effect from the date of its acceptance by the concerned Managing Committee or on the expiry of thirty days from the date of its receipt by the concerned Managing Committee, whichever is earlier.

Disqualifications

33. (1) An employee of the Government of India or any State Government or of a Local Authority or any institution receiving aid from

the funds of the Government, shall be disqualified for election or for continuing as a Chairperson or a Director of a Managing Committee of the Water Users' Association.

(2) No person who has been convicted by a criminal court for any offence involving moral turpitude shall be qualified for election or being or continuing in the office of a Chairperson or a Director of a Management Committee of a Water Users' Association.

(3) A person shall be disqualified for being elected as the Chairperson or a Director of the Managing Committee of a Water Users' Association if, on the date fixed for the scrutiny of nominations for election, or on the date of nominations, he is, (a) of unsound mind and stands so declared by a competent court; (b) an undischarged insolvent; (c) has defaulted in the payment of land revenue or water tax or charges payable either to the Government or to the Water Users' Association; (d) interested in a subsisting contract made with, or any work being done for, a Municipal Council or Corporation or the Panchayat Samiti or a Zilla Parishad or any State or Central Government or the Water Users' Association:

Provided that, a person shall not be deemed to have any interest in such contract or work by reason only of his having share or interest in,- (i) the company as a mere share-holder but not as a director; or (ii) any lease, sale or purchase of immovable property or any agreement for the same; or (iii) any agreement for the loan of money or any security for the payment or money only; or (iv) any newspaper in which an advertisement relating to the affairs of the Water Users' Association is inserted.

Explanation.– For the purposes of this clause, where a contract is fully performed it shall not be deemed to be subsisting merely on the ground that the concerned Municipal Council or Corporation or Panchayat Samiti or Zilla Parishad, or Water Users' Association, State or Central Government has not performed its part of the contractual obligations.

(4) A Chairperson or a Director of the Managing Committee of a Water Users Association shall be disqualified for election to or continuing in such office of such office bearer absents himself from three consecutive meeting of the Managing committee without reasonable cause and without prior intimation to the Committee in writing.

(5) A person having more than two children at a time of contesting the election of the Chairperson or a Director of the Managing Committee shall be disqualified for such election:

Provided that, a person having more than two children on the date of commencement of this Act, shall not be disqualified under this sub-section, for so long as the number of children he had on the date of such commencement, has not increased:

Provided further that, a child or more than one child born in at one time, within a period of one year from the date of such commencement shall not be taken into consideration for the purposes of disqualification under this section.

Explanation.– For the purposes of this sub-section,- (i) where a couple has only one child on or after the date of such commencement, any number of children born subsequently at one time only shall be deemed to be one child for the purposes of this section; (ii) "child" does not include an adopted child or children.

(6) A person shall become disqualified to continue as Director of Water Users' Association or hold the office of Chairperson or the Director of a Managing Committee of a Water Users' Association, if he ceases to be a landholder or occupier in the area of operation of the concerned Water Users' Association.

Filling up of Vacancies

34. (1) A vacancy arising either due to recalling of a Director under section 19 or disqualification under section 33 or due to death, resignation or by any other reason, shall be filled up by nomination in the prescribed manner.

(2) The term of office of a Director or a Chairperson of a Water Users' Association, nominated under sub-section (1), shall be co-terminus with the term of the Managing Committee.

Records

35. (1) Every Water Users' Association shall keep at its office, the following books of accounts and records namely:-(a) A register of members of the Water Users' Association, containing the names of the members, their holding of lands in the area of operation of the Water Users' Association and their Water Entitlements in the Normal Year;(b) copy of this Act duly modified up-to-date; (c) a map of the area of operation of the Water Users' Association alongwith map of the structures and canal systems prepared in consultation with the Irrigation Department of Government; (d) a statement of the assets and liabilities; (e) a register of the proceedings of the General Body meetings; (f) a register of the proceedings of the Managing Committee meetings; (g) a register of proceedings of the election and recall of Managing Committee Director; (h) books of account showing receipts and payments; (i) books

of account of all purchases and sales of goods by the Water Users' Association; (j) register of measurement books, level field books, work orders and the like; (k) copies of audit reports and enquiry reports; (l) Water Accounts; and (m) all such other accounts, records and documents as may be prescribed.

(2) The books of accounts and other records maintained by the Water Users' Associations shall, be open for information to the Members of such Association, subject to the rules being made in this behalf.

Model Regulations

36. (1) The state Government shall, publish model regulations, for guidance and adoption by the Water Users' Association and its Managing Committee laying down guidelines relating to the number of meetings to be held, the procedure and the quorum in the meeting etc.

(2) The Water Users' Association shall adopt the model regulations, with necessary modification, if any, required, after recording reasons necessitating such modification, so however as not be entirely defeating its nature and purpose.

Dissolution of Managing Committee of Water Users' Association and Transitional Arrangements

37. (1) In case of any abuse of powers and failure to carry out functions or on the part of the Managing Committee of a Water Users' Association, or contravention of the provisions of this Act or Rules made thereunder, the concerned Appropriate Authority may, after giving the defaulting Managing Committee a reasonable opportunity to show cause, dissolve such Managing committee or pass such orders as deemed necessary; Provided that, on such dissolution, the Managing Committee shall be constituted within a period of three months from the date of dissolution.

(2) The State Government may, by notification in the Official Gazette, pending the reconstitution of the Managing Committee of a Water Users' Association appoint an officer, as prescribed to exercise the powers and perform the functions of such Managing committee.

(3) If the WUA does not function even after such reconstitution under sub-section (1), the Appropriate Authority may, after giving the Managing Committee of such association a reasonable opportunity of being heard take such necessary action, as may be prescribed.

Appointment of Competent Authority

38. (1) The Appropriate Authority may by notification, in the

Official Gazette, appoint or designate such number of officers of the State Government in the Irrigation Department, as it may deem fit, to be the Competent Authority in respect of any Water Users' Association for the purpose of this Act.

(2) The Competent Authority so designated under sub- section (1) shall provide technical advice and ensure that work is executed in accordance with technical parameters.

Chapter - III

LIFT IRRIGATION WATER USERS' ASSOCIATION

Water Users' Association for Lift Irrigation

39. (1) For every lift irrigation scheme under this Act, there shall be constituted a Water Users' Association.

(2) The State Government shall, by notification in the Official Gazette declare a cut-off-date, being a date not earlier than one year from the date of such notification, after which all the sanctions to individual schemes of lift irrigation shall stand cancelled in the delineated areas:

Provided that, the State Government may, by notification in the Official Gazette, extend the said period. Provided further that, under exceptional circumstances and having regard to geographical situations, the permission to individual lifts may be continued till the constitution of Cooperative Lift Irrigation Association.

Permission to form Lift Irrigation Water Users' Associations on Notified Rivers or Streams in Command of Flow Irrigation

40. In order to avoid overlapping in command areas of flow and lift irrigation and complications arising thereby, the Appropriate Authority may give permission to form Lift Irrigation Water Users' Associations, on notified rivers or streams in the command of flow irrigation, after delineating the area of flow and lift irrigation separately in prescribed manner.

Permission to Individual Lifts on Notified Rivers or Streams Upstream of Reservoir Beyond the Area of Operation of Lift Irrigation Water Users' Associations on Reservoir

41. In the case of Lift Irrigation on notified rivers or streams upstream of reservoir which are beyond the area of operation of Lift Irrigation Water Users' Associations or along the periphery of a reservoir

within the prescribed extent of water use and to all concerned project affected people as, prescribed, the Canal Officer may give permission to individual lifts and exercise control thereon:

Provided that, the Canal Officer shall, on constitution of the Lift Irrigation Water Users' Associations, cancel the individual sanctions given under this section:

Provided further that, the quantum of water to be allowed for lift irrigation along reservoir shall be determined in the prescribed manner depending upon the geographical situation and water availability.

Delineation of Areas of Operation of Lift Irrigation Water Users' Associations

42. (1) A Canal Officer not below the rank of an Executive Engineer, duly empowered by the Appropriate Authority in this behalf shall, by notification in Official Gazette and in accordance with the Rules made under this Act, delineate command areas of Lift Irrigation Water Users' Associations, separately based on the prescribed guidelines and declare those areas to be the areas of operation of respective Lift Irrigation Water Users' Associations for the purposes of this Act.

(2) The provisions of sub-sections (3), (4) (5) and (6) of section 6 shall, Mutatis Mutandis, apply to delineation of area of operation of Lift Irrigation Water Users' Associations.

Constitution of Lift Irrigation Water Users' Associations

43. (1) In the area delineated as an area of operation of the Lift Irrigation Water Users' Associations under section 42, Lift Irrigation Water Users' Associations shall be constituted by the holders or occupiers of lands so delineated. Such Association shall be registered in a prescribed manner.

(2) The provisions of sub-section (2) of section 8 shall, Mutatis Mutandis, apply to the constitution of Lift irrigation Water Users' Associations.

(3) The State Government may, for reasons to be recorded in writing, provide for any other type or form of Lift Irrigation Water Users' Association.

Managing Committee of Lift Irrigation Water Users' Association and Election of its Directors and Chairperson

44. The provisions of section 9 shall, Mutatis Mutandis, apply to

the Managing Committee of the Lift Irrigation Water Users' Association and election of its Chairperson and Directors.

Water Supply to all Lift Irrigation Water Users' Associations and individual Lifts to be Regulated by the Canal Officer

45. Water supply to Lift Irrigation Water Users' Associations and individual lifts in the State, if any, shall be controlled and regulated by the Canal Officer duly empowered by the Appropriate Authority in this behalf. The Canal Officer shall assist, in obtaining permission to install pumps in the course of a river: Provided that, if there is any difficulty in constitution of the Lift Irrigation Water Users' Associations on Main Canal, the Appropriate Authority may, after studying the feasibility, by order, hand-over the responsibility of individual lifts on uncommon side of main canal to nearby Water Users' Association, at minor level for a specified period:

Provided also that, the consent of the concerned Water Users' Association and the concerned holder or occupier of land shall be obtained for such handing over.

Lift Irrigation Water Users' Associations not to Lift Water Directly from the Main Canal, Branch Canal and Distributary

46. The Lift Irrigation Water Users' Associations shall not lift water directly from the Main Canal or Branch Canal or Distributary. Water shall be made available to Lift Irrigation Water Users' Association by flow under gravity through an arrangement in the in-take well built by the Lift Irrigation Water Users' Associations at their own cost. Provided that, in case of the existing, individual Lift Irrigation Schemes joining the Lift Irrigation Water Users Association, as well as the existing Lift Irrigation Water Users Associations, even after registration of such Associations under section 39, the condition of constructing the intake wells shall be relaxed.

Provision and Maintenance of Water Meters by Lift Irrigation Water Users' Associations

47. (1) Lift Irrigation Water Users' Associations shall provide, install, maintain and calibrate water meters for flow measurement at their cost:

Provided that, for every water lifting device, there shall be a separate water meter certified by the Bureau of Indian Standards (ISI)

(2) The Canal Officer, duly empowered in this behalf, shall have the power to inspect and test the water meters and pass such order he deems fit.

Regulation of Supply of Electricity to Lift Irrigation Water Users' Association. Mah. XXX VIII of 1976.

48. The Appropriate Authority may make separate arrangements for the control and regulation of supply of electricity to the Lift Irrigation Water Users' Associations and individual lifts as prescribed. The provisions of section 97 of Maharashtra Irrigation Act, 1976 shall, Mutatis Mutandis, apply to Lift Irrigation Water Users' Associations under this Act. The Canal Officer shall assist in obtaining the connection of electric power supply to the Lift Irrigation Schemes.

Determining Water Rates for Lift Irrigation Water Users' Association

49. The Appropriate Authority shall, in accordance with the rules made in this behalf, determine the separate water rates on volumetric basis for Lift Irrigation Water Users' Associations, considering the fact that all expenditure for installation and organisation and maintenance of Lift Irrigation scheme is borne by Lift Irrigation Water Users' Associations.

Application of Provisions of Water Users' Association

50. Save as otherwise provided by or under any provisions of this Act, the provisions relating to the Water Users' Association shall, Mutatis Mutandis, apply to the Lift Irrigation Water Users' Association.

Linkages of Lift Irrigation Water Users' Associations

51. (1) Lift Irrigation Water Users' Association on Main Canal shall be deemed to be the members the Canal Level Association.

(2) The Lift Irrigation Water Users' Association on reservoirs shall be deemed to be the members of Project Level Association.

Chapter – IV

POWERS AND FUNCTIONS OF WATER USERS' ASSOCIATIONS

Powers and Functions of Water Users' Association

52. (1) Every Water Users' Association at any minor level shall

perform the following functions namely :- (a) to seek the following information from the upper Level Association or the concerned Canal Officer, as the case may be, and maintain its record and pass it on to its members, (i) Applicable water use Entitlement of such Association and that of its members; (ii) Number of rotations planned; (iii) Irrigation interval fixed;

(b) to collect, check and either sanction or reject partly or fully water applications or water indents of members based on criteria prescribed and unless previous dues are paid no water shall be supplied to such members:

(c) to seek information regarding canal operation schedule from upper Level Association or the concerned Canal Officer, prepare detailed water distribution programme or Rotational Water Supply of members before every rotation and ensure volumetric supply to each members as per their entitlement; Provided that, in the case of Minor Irrigation Project having only one Water Users' Association, that Water Users' Association itself shall prepare canal operation schedule for the project;

(d) to convene a meeting of the Managing Committee before each irrigation season to guide and help the members regarding, (i) Canal operation schedule and water distribution programme; (ii) public notice to be issued; (iii) maintenance of canal system before commencement of season; (iv) the information about the latest decisions taken by the upper Level Associations and the Appropriate Authority;

(e) to receive applicable water entitlement from upper Level Association or the Canal Officer and supply it to the members as per their Water Use Entitlement strictly following principle of tail to head;

(f) to maintain rotation-wise or season-wise Water Account in the prescribed form to remit to the Canal Officer, the water charges within the prescribed period against the bills received from the Canal Officer :

(g) to regulate and monitor water distribution to the members;

(h) to assess water charges for the members and send bills in the prescribed form;

(i) to collect prescribed service charges for operation and maintenance from the lower Water Users' Association at minor level and to remit its share to the upper Level Association, or, the concerned canal officer, as the case may be;

(j) to carry out annual maintenance and repairs to canal system falling under its jurisdiction;

(k) to ensure that the members maintain their canal system in good condition and they receive agreed maintenance amount in time;

(l) to maintain the records as prescribed;

(m) to resolve disputes, if any, amongst the members;

(n) to raise and utilise resources mentioned in Chapter V of this Act;

(o) to undertake all allied activities related to irrigated agriculture;

(p) to prepare and submit annual report including the Water Audit and Audited Accounts to the General Body;

(q) to enable the Government to publish the Annual Irrigation Status Report, furnish to the canal Officer, the requisite information as prescribed in time

(r) to undertake any other activities required to accomplish the objectives of Water Users' Association.

(2) Without prejudice to the generality of the provisions of subsection (1), the Water Users' Association at minor level shall also perform the following functions on its area of operation :–

(a) to convene a meeting of the members before each irrigation season to share with them information available regarding water availability and water entitlement and invite their suggestions, if any, for efficient management of available water;

(b) to recover current water charges and previous dues, if any, from the members;

(c) to issue irrigation passes to members whose water applications are sanctioned and unless previous dues are paid, no membership shall be allowed ;

(d) to prepare water distribution programme or Rotational Water Supply (RWS) for sub-minors, outlets and amongst members within the limits set by canal operation schedule;

(e) to implement the Agreement.

(3) Without prejudice to the generality of the provisions of subsection (1), it shall also be the function of a Water Users' Association to send a copy of the bill of the water charges assessed under clause (h) of sub section (1) to the Canal Officer concerned.

(4) Without prejudice in the generality of the provisions of subsection (1), it shall also be the function of a Project Level Association to assist the Canal Officer in assessment of the water charges for non-irrigation

use, which are to be remitted by such users to the Canal Officer.

(5) The Project Level Association shall also perform the following functions, in addition to the functions under subsection (1), namely :-

(a) to prepare water budget or preliminary irrigation programme for project before commencement of each irrigation season and determine Applicable Water Entitlement, number of rotations and irrigation interval in consultation with the concerned Canal Officer;

(b) to assist the concerned Canal Officer in assessment of water charges for Lift Irrigation Water Users' Association on reservoirs and non-irrigation use, which are to be remitted by such users to the Canal Officer;

(c) to obtain and use area capacity table or curve of reservoir and discharge tables of head regulators of canals.

(6) Without prejudice to the generality of provisions of subsection (1), the Lift Irrigation Water Users' Association shall also perform the following functions, namely :-

(a) to give guidance to members regarding the crops and the areas to be irrigated in a season considering Applicable Water Use Entitlement, number of rotations and irrigation interval;

(b) to receive measured volume of water from the concerned Canal Officer and to supply it to the members.

(7) The Water Users' Association for giving its water quota by consent of all the members may enter into an annual tripartite (Water Users' Association, Canal Officer and Private bodies or Company) agreement as prescribed so as to enable the private body or company to ensure increased agricultural production for the common interest of the members in the command area.

Functions of Appropriate Authority and Canal Officer

53. The functions of the Appropriate Authority and powers and duties of Canal Officer shall be as follows :

(a) to delineate the areas of operation of Water Users' Associations at all levels and provide the information of the same to Water Users' Association ;

(b) to determine and notify Applicable Water Entitlement for Water Users Associations at all levels;

(c) to determine the method to arrive at Applicable Water Entitlement in deficit years;

(d) to ensure supply of water as per the Applicable Water Entitlement on bulk basis measured volumetrically to Water Users' Associations at all levels at agreed intervals every year;

(e) to rehabilitate, modernise, maintain and operate canal system not handed over to Water Users' Associations;

(f) to conduct joint inspection and ensure repairs and renovation works in consultation with Water Users' Association as prescribed;

(g) to enter into an Agreement with Water Users' Association;

(h) to rehabilitate canal system in delineated areas under the Management of Irrigation System by Farmers', before handing over such system to Water Users' Association;

(i) to provide necessary technical and managerial assistance to Water Users' Associations for their capacity building;

(j) to provide all information to Water Users' Associations regarding availability of water, canal operation schedules and any other information relevant to Operation and Management of water supply in the areas under Management of Irrigation System by Farmers';

(k) to issue after assessment in a prescribed manner, the bill for water supplied to a Water Users' Association and to recover the same ;

(l) without prejudice to the generality of the provisions of the section 53, the concerned Canal Officer shall, initiate constitution of Water Users' Association under Sections 10, 13 and 16 only after the minor level Water Users Associations (Primary Branch of Water Users' Association) has been operational in an efficient manner and the Appropriate Authority is satisfied in the manner prescribed.

(m) to carry out any other functions that may be prescribed to accomplish the objectives of Water Users' Associations.

Chapter - V

SOURCES OF FUNDS

Sources of Funds of Water Users' Association

54. The funds of the Water Users' Association shall be raised through the following resources,–

(i) Water Charges;

(ii) Interest on Deposits;

(iii) Deposits from the Members;

(iv) Borrowings;

(v) Donations;

(vi) Contributions by Members;

(vii) Grants;

(viii) Penalty and penal fees from Members; and

(ix) Fees for the services rendered.

Utilisation of Funds of Water Users' Association

55. The funds of Water Users' Association shall be utilised to fulfill the objects of the Water Users' Association in the prescribed manner.

Budget

56. In every financial year, the Managing Committee of a Water Users' Association shall prepare a budget in respect of the next financial year, showing the estimated receipts and expenditure of the committee and shall place the same before the General Body of the Water Users' Association for its approval in the prescribed manner.

Deposit and administration of the funds

57. The Water Users' Association shall deposit its funds in a Scheduled Bank or a Co-operative Bank or a Urban Bank or a Commercial Bank or a Post Office Savings Bank.

Reserve Fund

58. Every Water Users' Association shall maintain such "Reserve Fund" and utilise the same in the manner as prescribed.

Audit

59. Every Water Users' Association shall get its accounts audited annually in the prescribed manner.

Chapter – VI

OFFENCES AND PENALTIES

Offences and Penalties

60. (1) Any person who without any lawful cause, sanction or authority commits any of the Acts specified in sub-section (2) within the area operation of Water Users' Association shall be deemed to have

committed an offence under this Act and shall, on conviction, be punished with imprisonment which may extend to six months or with fine which may be upto ten times the annual water charges or with both.

(2) The acts referred to in sub-section (1) are as follows ;–

(a) damages, alters or obstructs canal system;

(b) interferes with the flow and water level in canal system;

(c) pollutes or fouls the canal water so as to diminish its utility for the given purposes so as to render it less fit for the given purpose;

(d) destroys, tampers with, or removes any apparatus or part of any apparatus meant for operation and maintenance of canal system;

(e) obstructs the implementation of scheduled irrigation;

(f) uses water in unauthorised or wasteful manner;

(g) lifts water in an unauthorised manner;

(h) endangers the stability of canal system;

(i) obstructs the flow in the drainage system;

(j) neglects the maintenance of canal system;

(k) prevents the lawful use of canal system by any person authorised to use the same;

(l) does not pay the water charges in time;

(m) contravenes any of the provisions of this Act or Rules made thereunder;

Compounding of Offences

61. (1) The Appropriate Authority may either before or after the institution of proceedings for any offence punishable under this Act, or the rules made thereunder, accept from any person charged with such offence, by way of composition thereof, a sum of money not exceeding one thousand rupees within such time as the Appropriate Authority may determine.

(2) On payment by such person of such sum as composition amount, such person, if in custody, shall be released from the custody and, if any proceedings in any criminal court have been instituted against such person in respect of an offence under this Act, the composition shall be deemed to be to an acquittal and no further criminal proceedings shall be taken against such person in respect of such offence.

Punishment under other Law not Barred

62. Nothing in this Act shall prevent any person from being

prosecuted and punished under any other law for the time being in force for any Act of omission or commission made punishable by or under this Act:

Provided that, no person shall be prosecuted and punished for the same offence more than once.

Chapter – VII

CONFLICT RESOLUTION

Settlement of Disputes

63. (1) Any dispute or difference concerned with the constitution, management, powers or functions of a Water Users' Association arising amongst members, shall be resolved by any Managing Committee of the Water Users' Association concerned.

(2) Any dispute or difference arising amongst a Member and the Managing Committee of a Water Users Association or between two or more Water Users Associations shall be resolved by the next immediate upper Level Association:

Mah. of 2005

Provided that, in case of dispute or difference arising between a member and the Managing Committee of a Project Level Association or between two or more Project Level Associations, such dispute shall be resolved by such Canal officer, as may be prescribed, whose decision thereon shall be [final and] binding on all concerned.

Mah. XXXVIII of 1976

(3) If the dispute is between the Water Users' Association and the appropriate Authority, such dispute shall be referred to such authority as may be prescribed in this behalf.

Explanation. – In cases where the Managing Committee of the upper Level Association at any level does not exist, then the dispute shall be resolved by the Managing Committee of next Upper Level Association or the concerned Canal Officer, as the case may be.

(4) Every dispute or difference under this section shall be disposed off within forty-five days from the date of reference of the dispute or difference.

(5) For resolving the dispute as regards water use entitlement and

water distribution, the provisions of section 21 of the Maharashtra Water Resources Regulatory Authority Act, 2005 shall apply and the decisions regarding compensation shall be taken in accordance with the provisions of section 75 of the Maharashtra Irrigation Act, 1976.

Appeals

64. (1) Any person aggrieved by any decision made or order passed by any Managing Committee of a Water Users Association at any level may appeal to the Managing Committee of the immediate Upper Level Association.

(2) Any person aggrieved by any decision made or order passed by the Managing Committee of a Project Level Association may appeal to such authority as may be prescribed, whose decision thereon shall be final.

(3) Any appeal under this section shall be made within thirty days from the date of receipt of communication of the decision or from the date of the receipt of the order by the aggrieved party, whichever is earlier: Provided that, the appellate authority may, for sufficient reasons to be recorded in writing, condone the delay in filing the appeal beyond thirty days.

(4) Every appeal under this section shall be disposed off within thirty days from the date of filing of the appeal. Provided that, when the appeal is not disposed off within the period specified, the appellate authority shall record the reasons in writing for the delay beyond the period of thirty days.

Explanation.– In case immediate Upper Level Association at any level does not exist, then the Appeal may be made to the next Upper Level Association or concerned Canal Officer, as the case may be.

Chapter – VIII

GENERAL PROVISIONS

Application of the Provisions of the Maharashtra Water Resources Regulatory Authority Act, 2005

65. The provisions of the sections 11 to 14 and section 22 of the Maharashtra Water Resources Regulatory Authority Act 2005 shall apply for implementing this Act and the rules made there-under.

Ownership and Control of Specified Items or Components to Vest with State Government

66. Unless otherwise decided by the Government, the ownership and the control of the following shall vest in Government, namely:-

(i) Reservoirs and Head works of any irrigation project;

(ii) Main rivers and their tributaries;

(iii) Government buildings, colonies and lands;

(iv) Government vehicles;

(v) Government telecommunication systems:

Provided that, the Government may, upon request being made in that behalf, permit Water Users' Associations to use the facilities mentioned at clauses (iii), (iv) and (v) in the prescribed manner, on payment of such charges as may be prescribed. The Government may depute its employees to Water Users' Associations, if such request is made by Water Users' Association.

Control of Canal System not Handed over to Water Users' Association

67.(1) Every canal system except the canal system handed over to the Water Users' Associations shall be controlled, regulated and maintained by the State Government. (2) During natural calamity or in exceptional circumstances the special repairs work in the area of the project, handed over to the Water Users' Associations shall be carried out, as prescribed, by the Appropriate Authority.

Water Budgeting At Project Level

68. (1) Water budgeting, that is to say, preparing the preliminary irrigation programme for the project, in consultation with the Canal Officer duly empowered in this behalf, shall be the responsibility of Project Level Association. Such Preliminary Irrigation Programme shall be based on priorities determined by the State Government from time to time.

Explanation - When a Project Level Association does not exist, the preparation of preliminary Irrigation programme shall be the responsibility of the concerned Canal Officer.

(2) The Applicable Water Use Entitlement shall be worked out every year or, every season as the case may be, considering practical situation and directives issued by the State Government.

State Government to be Lawful Owner of Canal System

69. The ownership of the canal system handed over to the Water Users' Association for the purposes of management thereof shall be with the State Government and the State Government shall have to take the power to take control of such system in the larger public interest.

Water Supply for Non Irrigation Purposes

70. The Appropriate Authority shall regulate the water supply for non-irrigation purposes from Reservoir, Main Canals, Branch Canals, Distributaries and rivers in such a manner as to protect and safeguard the Applicable Water Entitlement of Water Users' Associations at all levels.

Recovery of Water Charges Mah. XXX VIII of 1976

71. Sections 88 and 89 of the Maharashtra Irrigation Act, 1976 shall Mutatis Mutandis apply to recovery of water charges in areas under the Management of Irrigation Systems by Farmers.

Act to Apply to Existing Water Users' Associations

72. Any existing Water Users' Association in a delineated area constituted prior to the date of commencement of this Act, under any law for the time being in force, shall be deemed to have been constituted under the provisions of this Act :

Provided that, areas of operation of such Water Users' Associations shall be delineated in accordance with the relevant provisions of this Act and for that purpose the existing agreement, if any, shall accordingly stand modified.

73. (1) This Act shall apply to ongoing projects from the date of its publication in the Official Gazette.

(2) The work of distribution system having discharge capacity not exceeding one cubic meter per second shall be carried out only after constituting Minor Level Water Users' Association under sections 5, 6, 7 and 8 of this Act, in the command area with the participation of Minor Level Water Users' Association as prescribed and area of operation shall be handed over to such Associations. As the minor level Water Users Associations are participated from construction stage the provisions under section 22 regarding repairs and renovations shall not apply to such Associations.

Authentication of Orders and Documents of Water Users' Association

74. (1) All permissions, orders, decisions, notices and other documents, except the Agreement, of the Water Users' Association shall be authenticated by the Chairperson of the Water Users' Association or Member-Secretary authorised by the Water Users' Association in this behalf.

(2) The Agreement, shall be signed by the Chairperson of the Water Users' Associations.

Actions not to be Invalidated by Infirmity or Vacancy, etc.

75. No Actions or proceedings of the Water Users' Association shall be invalid for the reason merely of the existence of any vacancy in, or defect in the constitution of such Association.

Rules

76. (1) The State Government may, by notification in the Official Gazette, make Rules for carrying out the purposes of this Act.

(2) Except when rules are made for the first time, all rules made under this Act shall be subject to the condition of previous publication.

(3) Every rule made under this Act shall be laid as soon as may be, after it is made, before each House of the State Legislature while it is in session for a total period of thirty days which may be comprised in one session or in two or more successive sessions, and if, before the expiry of the session in which it is so laid or the session or sessions immediately following, both Houses agree in making any modification in the rule or both Houses agree that the rule should not be made, and notify such decision in the Official Gazette, the rule shall from the date of publication of such notification have effect only in such modified form or be of no effect, as the case may be; so, however, that any such modification or annulment shall be without prejudice to the validity of anything previously done or omitted to be done under that rule.

Repeal Mah. XXX VII of 1976

77. On the commencement of this Act, in relation to the areas under the Management of Irrigation Systems by Farmers, [Subsection (1) and (2) of section 46,] sections 46 to 48, section 55,sections 57, 58, 60 and 61 to 74 of the Maharashtra Irrigation Act, 1976, shall be deemed to have been repealed: Provided that, the repeal shall not affect -

(i) the previous operation of any sections so repealed or anything duly done or suffered thereunder; or

(ii) any right, privilege, obligation or liability acquired, accrued or incurred under any section so repealed; or

(iii) any penalty, forfeiture or punishment incurred in respect of any offence committed against any section so repealed; or

(iv) any investigation, proceeding, legal proceeding or remedy in respect of any right, privilege, obligation, liability, penalty, forfeiture or punishment as, aforesaid, and any such investigation, proceeding, legal proceeding or remedy may be instituted, continued or enforced and any such penalty, forfeiture or punishment may be imposed as if this Act had not been passed :

Provided further that, subject to the preceding proviso, anything done or any action taken (including any charges created, appointments, rules, notifications, orders, summons, notices, warrants and proclamations made or issued, authorities and powers conferred or vested, record-of-rights prepared or revised, canals or any water works or water courses or field-channels constructed, any supply of water made, water rates charged, agreements or contracts made, any taxes or fees levied, any compensation awarded, any labour obtained or supplied for emergency works of canals, any rights acquired or liabilities incurred, any suits instituted or proceeding taken or appeal made), under any section so repealed shall, in so far as such thing done or action taken is not inconsistent with the provisions of this Act, be deemed to have been done or taken under the corresponding provision of this Act; and shall continue to be in force accordingly unless and until superseded by anything done or any action taken under this Act.

Saving Mah. XXX VIII of 1976

78. The provisions of Maharashtra Irrigation Act, 1976 which have not been specifically repealed by this Act shall remain in force and shall Mutatis Mutandis apply to areas Management of Irrigation Systems by Farmers.

Act to have Over Riding Effect

79. Notwithstanding anything contained in any other law for the time being in force, the provisions of this Act shall have overriding effect, in respect of the areas of the Management of Irrigation System by Farmers.

Power to Remove Difficulty

80. (1) If any difficulty arises in giving effect to the provisions of this Act or as to the first constitution or reconstitution of any Water Users' Association after the commencement of this Act, the Appropriate Authority, as the occasion may require, by order published in the Official Gazette, may take necessary action for removing difficulties:

Provided that, no such order shall be issued after the expiry of a period of two years from such commencement.

(2) All orders made under sub-section (1) shall, as soon as may be, after they are made, be laid on the table of the both Houses of the State Legislature.

Repeal of Mah. Ordinance XXIII of 2004

81. The Maharashtra Management of Irrigation Systems by Farmers Ordinance, 2004 is hereby repealed.

References

Abbie, L.; J. Q. Harrison and J. M. Wall (1982), Economic Returns to Investment in Irrigation in India, World Bank Staff Working Papers No. 336, The World Bank, Washington, D.C., U.S.A.

Adhikari, Bhim and Jon C. Lovett (2006), "Institutions and Collective Action: Does Heterogeneity Matters in Community – Based Resource Management?", Journal of Development Studies, Vol. 42, No. 3, April, 426-445.

Agrawal, A. (2001), "Commons Property Institutions and Sustainable Governance of Resources", World Development, Vol. 29, No. 10, pp. 1649-72.

Baland, J. M. and J. P. Platteau (1997), "Wealth Inequality and Efficiency in the Commons - Part I: The Unregulated Case", Oxford Economic Papers, Vol. 49, No. 4, pp. 451-482.

Baland, J. M. and J. P. Platteau (1998), "Wealth Inequality and Efficiency in the Commons - Part II: The Unregulated Case", Oxford Economic Papers, Vol. 50, No. 1, pp. 1-22.

Baland, Jean-Marie and Jean-Philippe Platteau (1999), "The Ambiguous Impact of Inequality on Local Resource Management", World Development, Vol. 27, No. 5, pp. 773-88.

Bardan, P. (2000), "Irrigation and Cooperation: An Empirical Analysis of 48 Irrigation Communities in South India", Economic Development and Cultural Change, Vol. 48, No. 4, pp. 847-65.

Bardhan, P. (1993), "Analytics of the Institutions of Informal Cooperation in Rural Development", World Development, Vol. 21, No. 4, April, pp. 633-39.

Bardhan, P. (1993), "Symposium on Management of Local Commons", Journal of Economic Perspectives, Vol. 7, No. 4, pp. 87-92.

Bardhan, P. and Dayton-Johnshon, J. (2002), "Unequal Irrigators: Heterogeneity and Commons Management in Large-Scale Multivariate Research", in E. Osterm; T. Dietz; N. Doisak; P.C. Stren; S. Stonich and E.V. Weber (eds.) Drama of the Commons,

The National Academic Press, Washington, USA, pp. 87-112.

Bardhan, Pranab and Isha Ray (2006), "Methodological Approaches to the Questions of the Commons", Economic Development and Cultural Change, Vol. 54, pp. 655-676.

Belsare, Sanjay (2001), "Participatory Irrigation Management in Katepurna Irrigation Project: A Success Story", (www.icid.org/belsare_2001.pdf).

Berg, Marrait Van Den and Ruerd Ruben (2006), "Small-Scale Irrigation and Income Distribution in Ethiopia", Journal of Development Studies, Vol. 42, No. 5, pp. 868-880.

Bhattarai, M. and A. Narayanamoorthy (2003), "Impact of Irrigation on Rural Poverty: An Aggregate Panel-data Analysis for India", Water Policy, Vol. 5, No. 5-6, pp. 443-458.

Brewer, R; S. Kolavalli; A.H. Kalro; G. Naik; S. Ramnarayan; K.V. Raju and R. Sakthivadivel (1999), Irrigation Management Transfer in India: Policies, processes and Performance, Oxford and IBH Publishing Co. Pvt. Ltd., New Delhi.

Chambers, Robert (1988), Managing Canal Irrigation, Oxford and IBH Publishing Co. Pvt. Ltd., New delhi.

CMIE (2006), Agriculture, Centre for Monitoring Indian Economy Pvt. Ltd., Mumbai, March.

CWC (1998, 2000 and 2004), Water and Related Statistics, Central Water Commission, Ministry of Water Resources, Government of India, New Delhi.

Dayton-Johnson, J. (2000), "Determinants of Collective Action on the Local Commons: A Model with Evidence from Mexico", Journal of Development Economics, Vol. 62, pp. 181-208.

Dayton-Johnson, J. and P. Bardhan (2000), "Inequality and Conservation on the Local Commons: A Theoretical Exercise", The Economic Journal, Vol. 112, pp. 577-602.

Deshpande R. S and A. Narayanamoorthy (1999), "An Appraisal of Watershed Development Programme Across Regions in India", Artha Vijnana, Vol. 12, No. 4, pp. 315-415.

Deshpande R. S. and A. Narayanamoorthy (2001) "Issues before the Second Irrigation Commission of Maharashtra", Economic and Political Weekly, Vol. 36, No.12, pp.1034-1043.

Deshpande R. S. and A. Narayanamoorthy (2002), "Policy for Agriculture and Rural Development in the Context of Tenth plan", Agricultural Situation in India, Vol. 59, No.6, September, pp. 363-374.

Deshpande, R. S and V. Ratna Reddy (1990), "Social Dynamics and

Farmers' Society: A Case Study of Pani Panchayat", Indian Journal of Agricultural Economics, Vol. 45, No. 3, pp. 355-361.

Deshpande, R. S and V. Ratna Reddy (1991), Watershed Development Approach in Fragile Resource Region: An Analytical Study of Maharashtra, Mimeograph Series No. 33, Gokhale Institute of Politics and Economics, Pune, Maharashtra.

Dhawan, B. D. (1988), Irrigation in India's Agricultural Development: Productivity, Stability, Equity, Sage Publications India Pvt. Ltd., New Delhi.

Easter, K. W. (2000), "Asia's Irrigation Management in Transition: A Paradigm Shift Faces High Transaction Costs", Review of Agricultural Economics, Vol. 22, No. 2, pp. 370-388.

FAI (various years), Fertiliser Statistics, Fertilisers Association of India, New Delhi

Gadgil, D. R (1948), Economic Effects of Irrigation: Report of a Survey of the Direct and Indirect Benefits of the Godavari and Pravara Canals, Publication No. 17, Gokhale Institute of Politics and Economics, Pune, Maharashtra.

Gandhi, P. R. (1992), History and Practice of Management of Irrigation Waters in Maharashtra, Water and Land Management Institute, Aurangabad (Maharashtra), Publication No. 1, July.

Gandhi, Vasant P. (1996), "Investment Behaviour in Indian Agriculture", Indian Journal of Agricultural Economics, Vol. 51, No. 4, October-December, pp. 543-559.

GOI (1972), Report of the Second Irrigation Commission, Ministry of Irrigation and Power, Government of India, New Delhi.

GOI (1992), Report of the Committee on Pricing of Irrigation Water, Planning Commission, Government of India, New Delhi.

GOI (2000), National Agricultural Policy, Ministry of Agriculture, Government of India.

GOI (2001), Report of the Working Group on Watershed Development, Rainfed Farming and Natural Resource Management for the Tenth Five Year Plan, Planning Commission, Government of India, New Delhi.

GOI (2002), Tenth Five Year Plan: 2002-2007, Vol. II, Planning Commission, Government of India, New Delhi.

GOI (various years), Area and Production of Principal Crops in India, Ministry of Agriculture, Government of India, New Delhi.

GOI (various years), Indian Agricultural Statistics, Ministry of Agriculture, Government of India, New Delhi.

GOM (1979), Committee to Study the Introduction of Eight Monthly Supply of Water from the Irrigation Projects in Maharashtra (Interim Report), February.

GOM (1984), Report of the Fact Finding Committee on Regional Imbalance, Government of Maharashtra, Planning Department, Mumbai.

GOM (1999), Maharashtra Water and Irrigation Commission Report, Government of Maharashtra, Mumbai.

GOM (2006), Economic Survey of Maharashtra 2005-06, Directorate of Economics and Statistics, Planning Department, Government of Maharashtra, Mumbai.

GOM (various years), Districtwise Agricultural Statistical Information, Part I and II, Commissionerate of Agriculture, Pune, Maharashtra.

GOM (various years), Economic Survey of Maharashtra, Directorate of Economics and Statistics, Planning Department, Government of Maharashtra, Mumbai.

Gulati, Ashok; Mark Svendsen; Nandini Roy Choudhury (1994), "Major and Medium Irrigation Schemes: Towards Better Financial Performance", Economic and Political Weekly, Vol. 29, No. 26, pp. A72-A79.

Gulati, Ashok; Ruth Meinzen-Dick and K.V. Raju (2005), Institutional Reforms in Indian Irrigation, International Food Policy Research Institute, Washington, D.C., and Sage Publications, New Delhi.

Hradin, R. (1982), Collective Action, John Hopkins University Press, Baltimore.

Hussain, I and M. A. Hanjra (2003), "Does Irrigation Water Matter for Rural Poverty Alleviation: Evidence from South and South-East Asia", Water Policy, Vol. 5, No. 5-6, pp. 429-442.

Kanbur, R. (1992), "Heterogeneity, Distribution and Cooperation in Commons Property Resource Management", Background Paper for the 1992 World Development Report, Washington, DC, The World Bank.

Keremane, G. P.; Jennifer Mckay and A. Narayanamoorthy (2006), "The Decline of Innovative Local Self-Governance Institutions for Water Management: the Case of Pani Panchayats", International Journal of Rural Management, Vol. 2, No. 1, pp. 107-122.

Kerr, John (2002), Watershed Development Projects in India: An Evaluation, Research Report 127, International Food Policy Research Institute, Washington, D.C., U.S.A.

Lam, W.F. (1998), Governing Irrigation Systems in Nepal: Institutions,

Infrastructure, and Collective Action, Oakland, California, ICS Press.

Lele, S. N. and R. K. Patil (1994), Farmer Participation in Irrigation Management: A Case Study of Maharashtra, Horizon India Books, New Delhi.

Mahendra Dev (1996), Agricultural Policy Framework for Maharashtra - Issues and Options, Proceeding/Projects Report, July, Indira Gandhi Institute of Development Research, Mumbai.

Maloney, C. T. and K. V. Raju (1994), Managing Irrigation Together: Practice and Policy in India, Sage Publication, New Delhi.

Marothia, D. K. (2003), "Enhancing Sustainable Management of Water Resources in Agricultural Sector: The Role of Institutions", Indian Journal of Agricultural Economics, Vol. 58, No. 3, pp. 406-27.

Mosse, David (1999), "Colonial and Contemporary Ideologies of Community Management: The Case of Tank Irrigation Development in South India", Modern Asian Studies, Vol. 33, No. 2, pp. 303-338.

MOWR (1987 and 2002), National Water Policy, Ministry of Water Resources, Government of India, New Delhi, September.

MOWR (1999), Report of the Working Group on Water Availability for Use, National Commission for Water Resources Development Plan, Ministry of Water Resources, Government of India, New Delhi.

Naik, G.; A. H. Kalro; J. B. Brewer; M. Samad and R. Sakthivadivel (2002), Assessing the Impact of Irrigation Management Transfer: Case Studies from Maharashtra, Oxford & IBH Publishing Company, New Delhi.

Narayanamoorthy, A. (1997), "Economic Viability of Drip Irrigation: An Empirical Analysis from Maharashtra", Indian Journal of Agricultural Economics, Vol. 52, No. 4, pp. 728-739.

Narayanamoorthy, A. (2001), "Irrigation and Rural Poverty Nexus: A Statewise Analysis", Indian Journal of Agricultural Economics, Vol. 56, No.1, January-March, pp. 40-56.

Narayanamoorthy, A. (2002), "Indian Irrigation: Five Decades of Development", Water Resources Journal, No.212, June, pp. 1-29.

Narayanamoorthy, A. (2003),"Averting Water Crisis by Drip Method of Irrigation: A Study of Two Water-Intensive Crops", Indian Journal of Agricultural Economics, Vol. 58, No. 3, July-September, pp. 427-437.

Narayanamoorthy, A. and M. A. Hanjra (2006), "Rural Infrastructure and Agricultural Output Linkages: A Study of 256 Indian Districts",

Indian Journal of Agricultural Economics, Vol. 61, No. 3, July-September, pp. 444-459.

Narayanamoorthy, A. and R.S. Deshpande (2003),"Irrigation Development and Agricultural Wages: An Analysis Across States", Economic and Political Weekly, Vol. 38, No. 35, pp. 3716-3722.

Narayanamoorthy, A. and R. S. Deshpande (2005), Where Water Seeps!: Towards a New Phase in India's Irrigation Reforms, Academic Foundation, New Delhi.

Narayanamoorthy, A. and S. S. Kalamkar (2004), Irrigation, a chapter prepared for the Maharashtra State Development Report, to be published by the Planning Commission, Government of India, New Delhi.

Olson, M. (1965), The Logic of Collective Action: Public Goods and the Theory of Groups, Cambridge Harvard University Press.

Ostrom, E. (1990), Governing the Commons: The Evolution of Institutions for Collective Action, Cambridge University Press, New York.

Ostrom, E. and Roy Gardner (1993), "Coping with Asymmetries in the Commons: Self-Governing Irrigation Systems Can Work", Journal of Economic Perspectives, Vol. 7, No. 4, pp. 93-112.

Ostrom, E; Roy Gardner and J. Walker (1993), Rules, Games and Common-Pool Resources, Ann Arbor, University of Michigan Press.

Pant, Niranjan (1999), "Impact of Irrigation Management Transfer in Maharashtra: An Assessment", Economic and Political Weekly, Vol. 34, No. 13, March 27, pp. A17- A26.

Parthasarathy, R. (2000), "Participatory Irrigation Management Programme in Gujarat: Institutional and Financial Issues", Economic and Political Weekly, Vol. 35, No. 35 and 36, August 26 - September 2, pp. 3147-3154.

Rajagopal, A.; R. Doraiswamy; Peter P. Mollinga; K. J. Joy and S. Paranjape (2002), Tailenders and Other Deprived in Irrigation in India: Tamil Nadu, Karnataka and Maharashtra, CWP Research Papers 14, Irrigation and Water Engineering, Wageningen University, The Netherlands.

Rao, V.M. (1992), "Change Processes in Dryland Communities", Indian Journal of Agricultural Economics, Vol. 47, No. 1, January-March, pp. 1-24.

Rath, Nilakantha (1989), "Agricultural Growth and Investment in India", Journal of Indian School of Political Economy, Vol. 1, No. 1, January-June, pp. 64-83.

Rath, Nilakantha (1997), A Note on Approach to Irrigation in Maharashtra, paper presented at the Workshop on Economic Problems of Water, organised by the Second Irrigation Commission of Maharashtra and Marathi Economic Association, Sangli, Maharashtra, May 14-15.

Rath, Nilakantha and A.K. Mitra (1989), "Economics of Irrigation in Water-Scarce Regions: A Study of Maharashtra", Artha Vijnana, Vol. 31, No. 1, pp.1-129.

Ray, Isha (2005), "Get the Price Right: Water Prices and Irrigation Efficiency", Economic and Political Weekly, Vol. 40, No. 33, August 13, pp. 3659-3668.

Reddy, M. A. (1996), "Kudimaramat" in B.C. Barah (ed.) Traditional Water Harvesting Systems: An Ecological Economic Survey, New Age International Publications, New Delhi, pp. 101-119.

Reddy, V. Ratna (2000), "Sustainable Watershed Management: Institutional Approach", Economic and Political Weekly, Vol. 35, No. 38, September 16, pp. 3435-3444.

Reddy, V. Ratna and P. Prudhvikar Reddy (2005), "How Participatory is Participatory Irrigation Management? Water Users' Associations in Andhra Pradesh", Economic and Political Weekly, Vol. 40, No. 53, December 31, pp. 5587-5595.

Rosegrant, W. Mark (1997), Water Resources in the Twenty-First Century: Challenges and Implications for Action, Food and Agriculture, and the Environment Discussion Paper 20, International Food Policy Research Institute, Washington D.C., U.S.A.

Sakurai, T. and K. Palanisami (2001), "Tank Irrigation Management as a Local Common Property: The Case of Tamil Nadu, India", Agricultural Economics, Vol. 25, pp. 273-283.

Saleth, R. M. (1996), Water Institutions in India: Economics, Law and Policy, Commonwealth Publishers, New Delhi.

Saleth, R. M.; M. Samad; D. Molden and I. Hussain (2003), "Water, poverty and Gender: A Review of Issues and Policies", Water Policy, Vol. 5, no. 5-6, pp. 385-398.

Saleth, R. Maria and Ariel Dinar (2004), The Institutional Economics of Water: A Cross Country Analysis of Institutions and Performance, Edward Elgar, United Kingdom.

Sawant, S.D.; B.N. Kulkarni; C.V. Achuthan and K. J. S. Satyasai (1999), Agricultural Development in Maharashtra - Problems and Prospects, Occasional Paper 7, National Bank for Agriculture and Rural Development, Mumbai.

Seabright, Paul (1993), "Managing Local Commons: Theoretical Issues in Incentive Design", Journal of Economic Perspectives, Vol. 7, No. 4, pp. 113-134.

Shah, T. and O. P. Singh (2004), "Irrigation Development and Rural Póverty in Gujarat, India: A Disaggregated Analysis", Water International, Vol. 29, No. 2, pp. 167-177.

Tang, S. Y. (1991), "Institutional Arrangements and the Management of Common-pool Resources", Public Administration Review, Vol. 5, pp. 42-51.

Tang, S. Y. (1992), Institutions and Collective Action: Self-Governance in Irrigation, San Francisco, ICS press.

Thakur, Manish, K and Binay K. Pattnaik (2002), "How Effective are 'Pani Panchayats'?: A Field View from Maharashtra", Sociological Bulletin, Vol. 51, No.2, pp. 243-268.

Vaidyanathan, A. (1994), "Transferring Irrigation Management to Farmers", Economic and Political Weekly, Vol. 29, No. 47, pp. 2965-2967.

Vaidyanathan, A. (1999), Water Resources Management: Institutions and Irrigation Development in India, Oxford University Press, New Delhi.

Vaidyanathan, A. (2001), Tanks in South India, Centre for Science and Environment, New Delhi.

Vaidyanathan, A.; A. Krishna Kumar; A. Rajagopal and D. Varatharajan (1994), "Impact of Irrigation on Productivity of Land", Journal of Indian School of Political Economy, Vol. 6, No.4, October-December, pp. 601-645.

Velded, T. (2000), "Village Politics: Heterogeneity, Leadership and Collective Action", Journal of Development Studies, Vol. 36, No. 5, pp. 105-34.

Vermillion, Douglas, L (1997), Impacts of Irrigation Management Transfer: A Review of the Evidence, Research Report 11, International Irrigation Management Institute, Colombo, Sri Lanka.

Wade, Robert (1979), "The Social Response to Irrigation: An Indian Case Study", The Journal of Development Studies, Vol. 16, No. 1, pp. 3-26.

Wade, Robert (1981), "The System of Administrative and Political Corruption: Canal Irrigation in South India", The Journal of Development Studies, Vol. 18, No. 3, pp. 287-328.

Wade, Robert (1981), "The World Bank and India's Irrigation Reform", The Journal of Development Studies, Vol. 18, No. 2, pp. 171-184.

Wade, Robert (1987), Village Republics: Economic Conditions for

Collective Action in South India, Cambridge University Press, Cambridge.

Wade, Robert (1995), "The Ecological Basis of Irrigation Institutions: East and South Asia", World Development, Vol. 23, No. 12, pp. 2041-2049.

World Bank (2002), INDIA, Maharashtra: Reorienting Government to Facilitate Growth and Reduce Poverty, Vol. I and II, Report No. 25053-IN, Poverty Reduction and Economic Management Unit, South Asia Region, The World Bank, Washington, D.C., USA.

World Bank (2005), India's Water Economy: Bracing for a Turbulent Future, Report No. 34750-IN, Agriculture and Rural Development Unit, South Asia Region, The World Bank, Washington, D.C., USA.

http://www.mahagovid.org

http://dird-pune.org/

http://www.walmi.org/

www.fao.org

www.sristi.org

http://waterconserve.maharashtra.gov.in

www.cgiar.org

www.mah.nic.in/sahakar/Nirmann.htm

INDEX